DESIGNING
MEANINGFUL
STEM
LESSONS

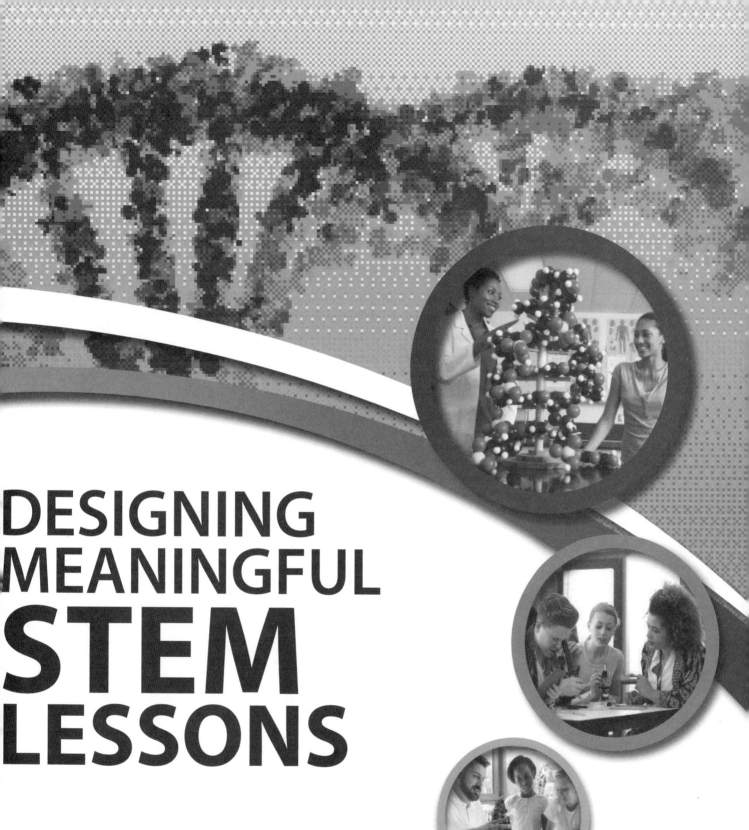

DESIGNING MEANINGFUL STEM LESSONS

Milton Huling and
Jackie Speake Dwyer

NSTApress

National Science Teachers Association

Arlington, Virginia

National Science Teachers Association

Claire Reinburg, Director
Rachel Ledbetter, Managing Editor
Deborah Siegel, Associate Editor
Andrea Silen, Associate Editor
Donna Yudkin, Book Acquisitions Manager

ART AND DESIGN
Will Thomas Jr., Director
cover and interior design

PRINTING AND PRODUCTION
Catherine Lorrain, Director

NATIONAL SCIENCE TEACHERS ASSOCIATION
David L. Evans, Executive Director

1840 Wilson Blvd., Arlington, VA 22201
www.nsta.org/store
For customer service inquiries, please call 800-277-5300.

NSTA is committed to publishing material that promotes the best in inquiry-based science education. However, conditions of actual use may vary, and the safety procedures and practices described in this book are intended to serve only as a guide. Additional precautionary measures may be required. NSTA and the authors do not warrant or represent that the procedures and practices in this book meet any safety code or standard of federal, state, or local regulations. NSTA and the authors disclaim any liability for personal injury or damage to property arising out of or relating to the use of this book, including any of the recommendations, instructions, or materials contained therein.

Library of Congress Cataloging-in-Publication Data
Names: Huling, Milton, 1961- author. | Dwyer, Jackie Speake, 1968- author.
Title: Designing meaningful STEM lessons / by Milton Huling and Jackie Speake Dwyer.
Description: Arlington, VA : National Science Teachers Association, [2018] | Includes bibliographical references and index.
Identifiers: LCCN 2017053280 | ISBN 9781681405568 (print) | ISBN 9781681405575 (e-book)
Subjects: LCSH: Science--Study and teaching--United States. | Technology--Study and teaching--United States. | Engineering--Study and teaching--United States. | Mathematics--Study and teaching--United States. | Curriculum planning--United States. | Instructional systems--United States--Design.
Classification: LCC Q181 .H935 2018 | DDC 507.1/2--dc23 LC record available at *https://lccn.loc.gov/2017053280*

CONTENTS

CONTENTS

FOREWORD

STEM—it seems to be in the air. It is featured in newspaper articles. Anchorpeople work the acronym into stories used to describe school programs. Policy makers take every opportunity to feature it when discussing schools. Meanwhile, scientists, engineers, mathematicians, and computer scientists are left scratching their collective heads when the acronym appears, as the notion seems to have emerged overnight, like a ring of mushrooms.

What is STEM, and what should it mean for education? Although the notion seems to be almost everywhere in the early part of the 21st century, there are very few serious treatments about what this notion is and what it may bring to the world of K–12 education. Indeed, it seems that anytime a discipline of science, mathematics, engineering, or technology is featured in a lesson, it is called STEM. So, what does this construct bring to education? Is it a passing political fad, an idea constructed as a rhetorical tool to draw attention to a project or lesson? Or is there something more to the construct?

Authors Milt Huling and Jackie Speake Dwyer weigh into this discussion, suggesting that STEM instruction must involve something more than simply including one or more STEM disciplines. In *Designing Meaningful STEM Lessons*, the authors approach STEM instruction from a science educator's point of view, arguing that in a STEM lesson, the "integration of math, technology and engineering should be used to support the learning of a science concept." To allow for this emphasis, Huling and Dwyer propose StEMT—an instructional model that begins with the well-known and well-established learning cycle and embeds an engineering challenge into the "elaborate" phase. This introduction is accompanied by a number of examples of "StEMTified" lessons to highlight the potential of this instructional model.

Designing Meaningful STEM Lessons brings clarity to the sometimes murky STEM conversations in ways that will be helpful to teachers and administrators alike. Informed by research findings and the practicalities of the work in schools and situated in national conversations around the *Next Generation Science Standards,* the StEMT instructional model and the activities Huling and Dwyer offer will be useful to a wide swath of K–12 science teachers.

Sherry A. Southerland
Editor, *Science Education*
Director, School of Teacher Education and FSU-Teach
Florida State University

PREFACE

As educators who are deeply involved in science instruction, we (the authors) understand the value of STEM as a focus in curriculum. As central office administrators, we work with and support STEM academies and STEM classes. Our district has embraced STEM just like every other school district in the nation. STEM curriculum as a core concept is extremely attractive—it embraces all of the curricula that will help our students succeed in life. School administrators and school boards easily understand that STEM is a means to engage and attract the best and brightest students. We have a good understanding of what goes into STEM and the things that you need for it, but the process of implementing STEM curriculum within the classroom is quite nebulous on the best of days. There is a dilemma for all educators and learners when crafting a response to the question, "What does STEM look like in the classroom?" The easy, truthful (and educationally unhelpful) response is that STEM is simply any configuration of the disciplines with little regard for the outcome of the lesson. In other words, anything can be called STEM as long as it incorporates one, some, or all of the four disciplines commonly associated with STEM.

Although the concept of STEM being whatever you need it to be sounds wonderful at a philosophical level, application of such a vague notion quickly becomes problematic when paired with the expectations of targeted and rigorous quality instructions. Building a race car is a great way for students to learn problem-solving skills, but in the absence of a framework for science education, it is just a cool project.

States have tried to address this gap by designing and adopting their own individual state standards for science, and the development of *A Framework for K–12 Science Education* (*Framework*; NRC 2012) has recently led to the integration of engineering practices as a way to enhance science instruction and to integrate the components of STEM, specifically the engineering component. The great thing about science standards in all states is that the big ideas are mostly consistent. It is how we teach and frame the content that makes a significant difference in student learning and retention of important science concepts. Standards are the framework, pedagogy is the method of delivering the content to students, and STEM provides the integral component of real-world application for learning. Many articles have been written describing STEM as a solution to problem solving or a way to help students become better problem solvers. For example, many references have been made to problem-based learning or project-based learning (PBL). Although these types of instructional methods are often mentioned, most articles focus on the integration of the components of STEM with little regard for how the instruction must occur.

It is here that our work begins with re-envisioning STEM. Within our method, research-based instructional methodologies are not abandoned; rather, they are embraced within the STEM framework that includes the relevance from the engineering component. In fact, our method infuses everything the science education community has learned through decades

PREFACE

of researching how children learn (e.g., Carey 1991, 1999; Chinn and Brewer 1993; Driver 1989; Driver et al. 1996; Posner et al. 198; Sinatra and Chinn 2011). Our goal was to design a simple, practical process for implementing meaningful STEM activities in the classroom. Our methodological process for doing STEM mirrors all we know about how students learn: It includes a problem for which the lesson is built, and it incorporates engineering practices and design processes into the sciences to enhance their relevance. We now have an intuitive foundation on which to build authentic intermediate elementary and middle school STEM lessons through the process of STEM that is not labor intensive and can be used with existing science lessons.

We have shared our vision at many trainings within our district as well as with the Florida Association of Science Supervisors. Just as in our first training, the participants were shocked at the simplicity of our concept, as it gave meaning to an otherwise nebulous concept. More recently, we were having a conversation with a colleague and the topic of STEM came up. We provided the Cliff Notes version on how STEM was a "process and not a thing" to better illustrate the connections of the disciplines and the flow of the process. The response: "That is the best explanation of how STEM should be taught that I have ever heard. You need to write a book for teachers." And so *Designing Meaningful STEM Lessons* was written. We hope elementary and middle school teachers enjoy the book and that it helps to bring STEM learning to life in their classrooms, easily and effectively.

Special Attributes of the Book

- Ready-to-use lessons targeting students in grades 4–8 are correlated to the *Framework* disciplinary core ideas (NRC 2012). According to the *Framework*, disciplinary core ideas should have **broad importance** across multiple sciences or engineering disciplines or be a **key organizing concept** of a single discipline; provide a **key tool** for understanding or investigating more complex ideas and solving problems; relate to the **interests and life experiences of students** or be connected to **societal or personal concerns** that require scientific or technological knowledge; and be **teachable** and **learnable** over multiple grades at increasing levels of depth and sophistication. In this book, disciplinary core ideas are grouped into three domains: the physical sciences; the life sciences; and Earth and space science. The fourth domain—engineering, technology, and applications of science—is embedded into the lessons and is not a separate domain.

- Standards-based instruction: Because science content is fairly consistent from state to state, this book will focus on the disciplinary core ideas as identified in the *Framework* (NRC 2012). The big ideas of science are consistent—it is how we teach this content that makes a significant difference in student learning and retention of important science concepts.

- Constructivism is a philosophy about learning that suggests that learners need to build their own understanding of new ideas. Two of the most prominent constructivist researchers are **Jean Piaget** (stages of cognitive development) and **Howard Gardner** (multiple intelligences).

- Inquiry: The National Science Education Standards (NRC 1996) define inquiry as

A set of interrelated processes by which scientists and students pose questions about the natural world and investigate phenomena; in doing so, students acquire knowledge and develop a rich understanding of concepts, principles, models, and theories. Inquiry is a critical component of a science program at all grade levels and in every domain of science, and designers of curricula and programs must be sure that the approach to content, as well as the teaching and assessment strategies, reflect the acquisition of scientific understanding through inquiry. Students then will learn science in a way that reflects how science actually works. (p. 214)

- *5E Instructional Model:* The Biological Sciences Curriculum Study (BSCS), a team led by Principal Investigator Roger Bybee, developed an instructional model for constructivism called the "Five Es." Although the BSCS 5Es and inquiry are not synonymous, the 5E Instructional Model is based on constructivist theories and enhances student inquiry through a series of planning strategies. It works well with StEMT lesson design, but is not the only model that works, because constructivism through any model is synonymous with good teaching.

- *Claims, Evidence, Reasoning (CER):* There is a plethora of research to support that learning content through processes are important in science, and when students construct scientific explanations, through discourse or writing, they learn important concepts (Krajcik and Sutherland 2009, 2010; McNeill and Krajcik 2009, 2012; NRC 1996, 2000, 2012). When writing scientific explanations, students apply scientific ideas to answer a question or solve a problem using evidence. Students construct scientific explanations that result in demonstrating mastery of a specific learning goal and ultimately, in life and the real world, must use evidence to communicate their ideas to other people. Engaging in scientific explanations improves students' ability to reason and become more adept at critical and analytical thinking (McNeill and Krajcik 2009, 2012). When students justify their claims with evidence and reasoning, teachers gain insight into students' thinking and understanding. By integrating the disciplinary core idea with the targeted engineering practice, a learning goal can be designed and formed into a guiding question that students can answer through discussion or by using claims, evidence, and reasoning. See Appendix A (p. 189) for a sample CER rubric.

PREFACE

References

Carey, S. 1985. *Conceptual change in childhood.* Cambridge, MA: MIT Press.

Carey, S. 1991. Knowledge acquisition: Enrichment or conceptual change? In *The epigenesis of mind,* ed. S. Carey and R. Gelman, 257–291. Hillsdale, NJ: Lawrence Erlbaum Associates.

Carey, S. 1999. Sources of conceptual change. In *Conceptual development: Piaget's legacy,* ed. E. K. Scholnick, K. Nelson, and P. Miller, 293–326. Hillsdale, NJ: Lawrence Erlbaum Associates.

Chinn, C. A., and W. F. Brewer. 1993. The role of anomalous data in knowledge acquisition: A theoretical framework and implications for science instruction. *Review of Educational Research* 63 (1): 1–49.

Driver, R. 1989. Students' conceptions and the learning of science. *International Journal of Science Education* 11 (5): 481–490.

Driver, R., J. Leach, R. Millar, and P. Scott. 1996. *Young people's images of science.* Bristol, PA: Open University Press.

Krajcik, J. S., and L. M. Sutherland. 2009. IQWST materials: Meeting the challenges of the 21st century. Paper presented at the National Research Council Workshop on Exploring the Intersection of Science Education and 21st Century Skills, Washington, DC.

Krajcik, J. S., and L. M. Sutherland. 2010. Supporting students in developing literacy in science. *Science* 328 (5977): 456–459.

McNeill, K. L., and J. S. Krajcik. 2009. *Claim, evidence, and reasoning: Supporting grade 5–8 students in constructing scientific explanations.* New York: Pearson.

McNeill, K. L., and J. S. Krajcik. 2012. *Supporting grade 5–8 students in constructing explanations in science: The claim, evidence, and reasoning framework for talk and writing.* Boston: Pearson.

National Research Council (NRC). 1996. *National science education standards.* Washington, DC: National Academies Press.

National Research Council (NRC). 2000. *Inquiry and the national science education standards: A guide for teaching and learning.* Washington, DC: National Academies Press.

National Research Council (NRC). 2012. *A framework for K–12 science education: Practices, crosscutting concepts, and core ideas.* Washington, DC: National Academies Press.

Posner, G. J., K. A. Strike, P. W. Hewson, and W. A. Gertzog. 1982. Accommodation of a scientific conception: Toward a theory of conceptual change. *Science Education* 66 (2): 211–227.

Sinatra, G. M., and C. A. Chinn. 2011. Thinking and reasoning in science: Promoting epistemic conceptual change. In *Educational psychology: Contributions to education.* Vol. 3, ed. K. Harris, C. B. McCormick, G. M. Sinatra, and J. Sweller. Washington, DC: American Psychological Association.

ACKNOWLEDGMENTS

A special thank you to the teachers from Polk County Public Schools, who tested the StEMT lessons in their classrooms and provided the feedback to continuously improve the StEMT lessons.

ABOUT THE AUTHORS

Dr. Milton Huling is a nerd … and proud of it. Milt loves modeling instruction in classrooms and making science come to life for students. Milt received a Bachelor of Science degree from Southern Illinois, Edwardsville, in Earth space/geology. He received a master's degree in science education from Florida State University along with the additional course work to become a certified teacher. Milt received a Ph.D. from the University of South Florida in curriculum and instruction with an emphasis in science education.

Milt's teaching career spans 14 years, starting in Illinois and continuing in Florida. He has been with the same Florida school district for 10 years. He has spent six of these years teaching physics and Earth science at the high school level. For the past four years, Milt has served as the District Secondary Science Coordinator and more recently as the District Elementary Curriculum Specialist.

On a daily basis, Milt designs curriculum and liaisons with the district's technology departments to deliver resources to both teachers and students. Milt also serves the school administration and teachers by developing and facilitating professional development opportunities.

Dr. Jackie Speake is a science geek … and proud of it. Jackie loves science and believes that everyone can learn through science exploration, even if they aren't a geek or a nerd. Science is for ALL! Jackie received her Bachelor of Science degree from the University of Maryland, College Park (go Terrapins!), in marine biology. After three years as a field biologist for the Maryland Department of Natural Resources, two years as a water chemistry lab technician at the National Aquarium in Baltimore, and three years as a field biologist with the Florida Department of Environmental Protection, Jackie decided to go back to school in 1997 to get her Master of Education degree in secondary science curriculum from the University of South Florida (USF) and became a certified high school biology teacher. As a lifelong learner who really enjoys collaborating and learning with colleagues, Jackie returned to USF to study education policy and leadership and received her doctorate in 2011.

Jackie's 20-year education career ranges from high school teacher; district curriculum specialist in a Florida district with 18,000 students; state of Florida Department of Education science program coordinator; and senior director for science in a large Florida school district with more than 100,000 students. She now works with a district science team to facilitate the development and implementation of curriculum, assessments, inquiry lessons, professional development, and community partnerships.

ABOUT OUR APPROACH

You might pick up this book because you are on the continuing search for ways to teach STEM, as we (the authors) have been on for many years. If you are looking for an effective way to teach STEM, you may be thinking, "I know what STEM is, so what is different about this approach?" In the following pages, we will answer those two important questions and give you step-by-step directions on how to use our approach to integrate STEM using our approach into your classroom instruction. You will also see the term "StEMT" pop up within various conversations within the book. You may wonder, What is StEMT? StEMT is our way of making sense of STEM as a process, which is really at the heart of our vision. We are definitely not out to challenge the acronym that has become so much a part of the educational vernacular. When you see this term within the book, it is to purposefully guide you to a new way of thinking about STEM as an instructional approach.

This is not a book that replaces integration or is juxtaposed to STEM. We are huge proponents of STEM education, and we believe in integration, but only if it is done in a meaningful way. We also believe that STEM is students learning the skills and mindset necessary to become the next generation of engineers, scientists, and mathematicians. Although STEM as a concept has become mantra-like within the educational landscape, teachers have nothing to guide them in gaining an understanding of what it looks like to be included in their classrooms on a daily basis. It is this problem for which we have a solution. That solution is StEMT. Again, please do not run out and change your school signage or even your stationery. StEMT is a way that helped us think about the process of a STEM lesson. StEMT as a process offers clarity on how to build the lens for teachers and students to use when approaching problems in a way that applies the skills and tools of STEM. StEMT offers a structured approach that provides a clear process for designing effective, meaningful STEM lessons. The seemingly simple reframing of how we think about STEM provides us with the tools to take the nebulous concept of STEM and transform it into a conceptual framework.

StEMT (or STEM if you still prefer) is about effective, relevant science instruction that provides real-world experiences for every child. Experiences with STEM must be available to all students. StEMT is *not* an instructional model; it is a conceptual framework, and its purpose is to help teachers effectively teach STEM concepts by providing a mechanism to link science and mathematics to activities that promote constructivist learning to allow students to be cognitively challenged through questioning and problem-solving opportunities. StEMT fits into research-based instructional models that already exist and is an integral part of good instruction. In fact, by StEMT*ifying* your lessons, you make the lesson both relevant and exciting for students. Our goal is to help teachers increase the efficacy of their current science lessons by substituting StEMT into a portion of their existing lessons. Before we get to lesson development, it is important to understand the creation of the StEMT conceptual framework and to do that, you must understand its origins.

ABOUT OUR APPROACH

The StEMT Idea

Almost every educator can describe a nightmare scenario in which he or she is providing professional training to peers and has inadvertently misspelled something in the presentation. Most people feel discomfort when someone we work with presents something that is misspelled in their work. With that as a frame, the first discussion of StEMT as an approach occurred during a professional development presentation given by Dr. Milton Huling. Picture a large die being extracted from the back of Milt's SUV. Its size was carefully matched to the cargo area of the vehicle. In fact, to ensure a perfect fit, it was constructed in the parking lot of a local Lowe's. Made of PVC pipe and wrapped with red-and-white plastic tablecloths, the sides were adorned with numbers. The cube was carried into the school and placed on the front table in the school's teacher training room. As Milt carried the large colorful die into the school, it drew everyone's attention, including that of his colleague, Dr. Jackie Speake.

The presentation began with the usual welcomes and introductions. Jackie looked up at the first slide in bewilderment. "Why does the first slide say, 'StEMT?' " she asked. Milt's explanation detailed that his recent readings had led him to recognize that to truly incorporate STEM into a typical science classroom, the whole concept needed to be re-envisioned. The only way to get this across to teachers in a professional development setting was to make it dynamic. If technology is the solution to a human problem and most papers about STEM mention problem solving as a component, it only makes sense that the solution to the problem must be the technology that is produced. *This is in conflict with the view that the "T" in STEM is all about integration of technology and not a solution to a problem.*

Dr. Huling's presentation itself included activities designed to parse out the difference between a scientific question and a question about the development of technology. The activity was drawn and modified from a Nature of Science activity called *The Cube* (Lederman and Abd-El-Khalik 2002) and the *Three Cube Method* (Lederman, n. d.). Once teachers have a conceptual understanding of the difference between science and technology, it is much easier to understand how math and engineering (the tools) are used to bridge the divide. Science provides the basic knowledge through inquiry. With understanding of scientific principles in hand, only then can we use, or apply, this knowledge toward the development of a technology (i.e., a solution to a problem). The mathematics and engineering are the means of getting from science understandings to technology.

S[T] Science infusing (raised through the power of) Technology (a.k.a. *Good Science*)

E Engineering design process

M Mathematical practices and habits of mind

T Technology as a product or solution to a problem

This migration from principle to application, we feel, is critical to provide the relevance to the learning of science that is often lacking within many science classrooms. StEMT as a process, as opposed to STEM as a thing, also has the potential to enrich the science knowledge of students. In this way, StEMT/STEM does not broaden the curriculum. Instead, it acts as a focus.

It was the initial reception by educators of the StEMT training at one elementary school that set the stage for an expansion of STEM trainings on the StEMT process. These trainings occurred at multiple levels: district trainings, state conferences, and mathematics and science partnership activities. Within our district, we use StEMT as the approach to teach STEM, which is logical, especially if we think in terms of problem-based learning.

The reframing of the approach and the technology for teachers has accentuated that the importance of technology is not how it is used within the lesson—it is more so that technology (or the solution to the problem) is the outcome of instruction. It is not to say that integrating technology is not critical because it is essential for students to be exposed to the use of technology. More important, however, is what question is answered as part of the StEMT lesson, and the designing of StEMT questions is critical to helping teachers implement STEM in their own classrooms.

All lessons have been adapted from multiple open education resource sites, and most of the base activities are those you will recognize that have been circulating for years and have been effective for many educators across the nation. More often than not, we adapted the StEMT activities from the lessons we design for Polk County public schools.

References

Lederman, N. (n. d.). An introduction to the nature of science and technology. Department of Mathematics and Science Education, Illinois Institute of Technology. *https://science.iit.edu/sites/science/files/elements/mse/pdfs/RR-ThreeCubeMethod-Norm.pdf.*

Lederman, N. G., and F. Abd-El-Khalick. 1998. Avoiding denatured science: Activities that promote understandings of the nature of science. In *The nature of science in science education: Rationales and strategies,* ed. W. McComas, 83–126. Dordrecht, Netherlands: Kluwer Academic Publishers.

CONNECTIONS TO STANDARDS

Lessons by Disciplinary Core Idea (DCI) From *A Framework for K–12 Science Education* (NRC 2012)

Chapter	Lesson Title	Grade	Disciplinary Core Idea and Component
7	To Infinity and Beyond	3–6	Earth and Space Sciences: Earth and the Solar System (ESS1.B)
	Oh, How the Garden Grows	3–6	Earth and Space Sciences: Earth Materials and Systems (ESS2.A)
	Separation Anxiety	3–6	Earth and Space Sciences: Natural Hazards (ESS3.B)
	The Human Factor	3–6	Earth and Space Sciences: Human Impacts on Earth Systems (ESS3.C)
8	Cell-fie	5–8	Life Sciences: Structure and Function (LS1.A)
	Itsy, Bitsy Spiders	5–8	Life Sciences: Interdependent Relationships in Ecosystems (LS2.A)
	Gobble, Gobble, Toil and Trouble	5–8	Life Sciences: Ecosystem Dynamics, Functioning, and Resilience (LS2.C)
	Designing the Super Explore	5–8	Life Sciences: Inheritance and Variation of Traits (LS3.A and LS3.B)
	It's a What?	5–8	Life Sciences: Natural Selection (LS4.B)
9	Is It Air or Wind?	3–6	Physical Sciences: Structure and Properties of Matter (PS1.A)
	Aircraft Catapult	3–6	Physical Sciences: Forces and Motion (PS2.A)
	Keepin' It Cool!	3–6	Physical Sciences: Conservation of Energy and Energy Transfer (PS3.B)
	The Pitch in the Wave	3–6	Physical Sciences: Wave Properties (PS4.A)
	Mirror, Mirror, on the Wall	3–6	Physical Sciences: Electromagnetic Radiation (PS4.B)
Integrated Into Lessons			Engineering design (ETS1) and links among engineering, technology, science, and society (ETS2)

CONNECTIONS TO STANDARDS

Standards for Mathematical Practice

The eight standards for mathematical practice as identified by the *Common Core State Standards for Mathematics* (NGAC and CCSSO 2010) are as follows:

1. Make sense of problems and persevere in solving them
2. Reason abstractly and quantitatively
3. Construct viable arguments and critique the reasoning of others
4. Model with mathematics
5. Use appropriate tools strategically
6. Attend to precision
7. Look for and make use of structure
8. Look for and express regularity in repeated reasoning

Science and Engineering Practices

The eight practices of science and engineering that *A Framework for K–12 Science Education* (NRC 2012) identifies as essential for all students to learn are as follows:

1. Asking questions (for science) and defining problems (for engineering)
2. Developing and using models
3. Planning and carrying out investigations
4. Analyzing and interpreting data
5. Using mathematics and computational thinking
6. Constructing explanations (for science) and designing solutions (for engineering)
7. Engaging in argument from evidence
8. Obtaining, evaluating, and communicating information

References

Lederman, N. (n. d.). An introduction to the nature of science and technology. Department of Mathematics and Science Education, Illinois Institute of Technology. *https://science.iit.edu/sites/ science/files/elements/mse/pdfs/RR-ThreeCubeMethod-Norm.pdf.*

Lederman, N. G., and F. Abd-El-Khalick. 1998. Avoiding denatured science: Activities that promote understandings of the nature of science. In *The nature of science in science education: Rationales and strategies,* ed. W. McComas, 83–126. Dordrecht, Netherlands: Kluwer Academic Publishers.

National Science Teachers Association

National Governors Association Center for Best Practices and Council of Chief State School Officers (NGAC and CCSSO). 2010. *Common Core State Standards.* Washington, DC: NGAC and CCSSO.

National Research Council (NRC). 2012. *A framework for K–12 science education: Practices, crosscutting concepts, and core ideas.* Washington, DC: National Academies Press.

U.S. Department of Education. 2015. *Mathematics and science partnerships. http://www2.ed.gov/ programs/mathsci/index.html.*

CHAPTER

1 | What Is STEM?

STEM (science, technology, engineering and mathematics) within the educational landscape has become cliché. The pervasiveness of STEM can be illustrated by the numerous organizations that now embrace it, including the Department of Homeland Security, the National Science Foundation, and the Boy Scouts of America (Boy Scouts of America 2013; DHS 2012; NSF 2010). How we rank nationally or internationally consumes most discussions about the state of affairs in education. The solution seems to always be that we need more emphasis on STEM, but what is STEM?

The first appearance of the acronym *STEM* appeared in the 1990s (Capraro et al. 2015) and has since spread like wildfire through the educational landscape. STEM can refer to anything that involves the four disciplines of science, mathematics, technology and engineering (Bybee 2010). Perhaps not surprisingly, the most natural integration of these disciplines can occur between mathematics and science, as mathematics grew from the discipline of science as a way to help explain the natural world using empirical means. Technology and engineering have been overshadowed by their STEM counterparts within the STEM movement. Although some technology teachers believe that they are doing their fair share by integrating technology and engineering, they still view STEM as four independent disciplines (Sanders 2009). Bybee (2010) argues that it is critical that students become familiar with technology. The use of technology is then critical to the application of science, engineering, and mathematics.

This vision of what STEM should look like has left many with more questions than answers. Foremost, what is STEM beyond its basic components? Is STEM any of the four included disciplines? Is it integration of some or all of the disciplines? The entrance of even more disciplines to the cadre of four leaves one to ponder how these subjects fit together (e.g., STEAM, STREAM, STREAMSS). Many believe that the inclusion of the arts is essential to STEM. Others make a case for reading and social studies. At what point do we stop adding letters to STEM and start calling it by its original name: school? Our response to those who want to include more letters to the STEM acronym is that we must first get our STEM house in order before we add more disciplines into the mix. The question remains: is STEM one thing or a continuum from the typical siloed courses to full-on integration of the components?

For some, STEM is portrayed as a classroom where students design and program robots. You would find few who would disagree that technology engages students and that these types of activities promote student collaboration. Nationally and internationally, there have been countless conferences on the topic of STEM. It would seem that every possible industry partner has something to offer in the way of STEM education. We seem to be left with a few underlying questions—namely, what is STEM and how will a teacher infuse the concept of STEM into his or her classroom? It is the latter that seems to be the most elusive, given that most teachers cannot stop teaching biology or physics and instead teach programming or some other technology-based skill to their students.

Even with the release of the *Next Generation Science Standards,* recognizing the similarity between the STEM initiatives all too common to most of us and the desired outcomes described in the *Framework for K–12 Science Education (Framework;* NRC, 2012), *classroom application of meaningful STEM activities and lessons remains elusive.* With technology being infused into every portion of our lives at an ever-increasing rate, how will education keep pace? The answer to that question may be that it may not—particularly if the educational community continues to insist on providing our students with skills that may be obsolete even before they graduate. Instead, we should be preparing students with the knowledge, skills, and tools to develop solutions for problems that don't yet exist (Darling-Hammond 2010). It's hard to see how integrating these four disciplines in itself can produce such lofty results.

The goal of this publication is to support teachers in their integration of STEM into their classroom, effectively and efficiently using all the research-based instructional strategies that teachers are familiar with. The chapters of this publication are aligned to (1) give the reader an overview of the STEM landscape; (2) provide a rationale for our conceptual framework; and (3) give step-by-step instructions on how you can StEMT*ify* a typical science lesson. More examples of lessons that illustrate just how easy it is to bring STEM into your classroom are also provided.

Safety Considerations for Hands-On Activities

Throughout this book, safety precautions are provided for classroom activities in the form of safety notes. Teachers should also review and follow local polices and protocols used within their school districts. Additional applicable standard operating procedures can be found in the National Science Teachers Association's *Safety in the Science Classroom, Laboratory, or Field Sites* (*http://www.nsta.org/docs/SafetyInTheScienceClassroomLabAndField.pdf*).

Disclaimer: The safety precautions of each activity are based in part on use of the recommended materials and instructions, legal safety standards, and better professional practices. Selection of alternative materials or procedures for these activities may jeopardize the level of safety and therefore is at the user's own risk.

References

Boy Scouts of America. 2013. STEM in scouting. *www.scouting.org/stem.aspx.*

Bybee, R. W. 2010. Advancing STEM education: A 2020 vision. *Technology and Engineering Teacher* 70 (1): 30–35.

Capraro, M. M., R. M. Capraro, S. Metoyer, S. Nite, and C. A. Peterson. 2015. Promising practices in STEM teaching and learning: A meta synthesis. Paper presented at the STEM Collaborative for Teacher Professional Development, Texas A&M University, College Station, TX.

Darling-Hammond, L. 2010. *The flat world and education: How America's commitment to equity will determine our future.* New York: Teachers College Press.

Department of Homeland Security (DHS). 2012. DHS announces expanded list of STEM degree programs. *www.dhs.gov/news/2012/05/11/dhs-announces-expanded-list-stem-degree-programs.*

National Research Council (NRC). 2012. *A framework for K–12 science education: Practices, crosscutting concepts, and core ideas.* Washington, DC: National Academies Press.

National Science Foundation (NSF). 2010. Preparing the next generation of STEM innovators: Identifying and developing our nation's human capital. *https://www.nsf.gov/nsb/publications/2010/nsb1033.pdf.*

Sanders, M. 2009. STEM, STEM education, STEMmania. *Technology Teacher* 68 (4): 20–26.

CHAPTER

2 | What Constitutes a STEM Program of Study?

The acronym *STEM* refers to science, technology, engineering, and mathematics. Workers in STEM occupations use science and mathematics to solve problems and drive our nation's innovation and competitiveness by generating new ideas, new companies, and new industries (Honey, Pearson, and Schweingruber 2014).

The Four STEM Disciplines

Science is the study of the natural world, including the laws of nature associated with physics, chemistry, and biology, as well as the treatment or application of facts, principles, concepts, or conventions associated with these disciplines (Morrison 2006). Science is a body of knowledge that has been accumulated over time and a process—scientific inquiry—that generates new knowledge. Knowledge from science informs the engineering design process (Figure 2.1).

Technology, although not a discipline in the strictest sense, comprises the entire system of people and organizations, knowledge, processes, and devices that go into creating and operating technological artifacts, as well as the artifacts themselves.

Figure 2.1. Engineering design process

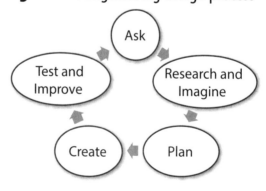

CHAPTER 2

Throughout history, humans have created technology to satisfy their wants and needs. Much of modern technology is a product of science and engineering, and technological tools are used in both fields (National Academy of Engineering and NRC 2002).

Engineering is both a body of knowledge—about the design and creation of human-made products—and a process for solving problems. This process is design under constraints. One constraint in engineering design is the laws of nature, or science. Other constraints include time, money, available materials, ergonomics, environmental regulations, manufacturability, and reparability. Engineering draws on concepts in science and mathematics as well as technological tools (National Academy of Engineering and NRC 2014).

Mathematics is the study of patterns and relationships among quantities, numbers, and space. Unlike in science, where empirical evidence is sought to warrant or overthrow claims, claims in mathematics are warranted through logical arguments based on foundational assumptions. The logical arguments themselves are part of mathematics, along with the claims. As in science, knowledge in mathematics continues to grow, but unlike in science, knowledge in mathematics is not overturned unless the foundational assumptions are transformed. Specific conceptual categories of K–12 mathematics include numbers and arithmetic, algebra, functions, geometry, and statistics and probability. Mathematics is used in science, engineering, and technology.

Most research has been focused on the cognitive outcomes of science and mathematics. These areas are of most interest as a means to gauge best practices in the current environment of accountability. The more social outcomes have relevance to choices for college majors, career choices, and subject aversion by groups of students. Cognitive outcomes, in the high-stakes testing environment, are needed to judge the health and effectiveness of STEM programs.

According to Janice Morrison (2006) from the Teaching Institute for Excellence in STEM (TIES), STEM programs should include the following features:

- A curriculum that is driven by problem solving, discovery, exploratory learning, and independent and collaborative research projects and requires students to actively engage in a problem involving a natural phenomenon to find a solution.

- A curriculum that incorporates the habits of mind for students to use technology; integrates engineering design; and requires systems thinking (collaboration and communication), maintenance, and troubleshooting.

- Innovative instruction that allows students to explore greater depths of all of the core subjects (English, mathematics, science, and social studies) by using the skills learned (reading, writing, speaking, listening, and computing).

- Technology that provides creative and innovative ways for students to solve problems and apply what they have learned conceptually.

What Constitutes a STEM Program of Study?

Schools integrating a STEM program into their curriculum must meet the needs of industry and higher education and focus on application and higher-level thinking. Students use the engineering design process to remove boundaries between core subjects, allowing for transfer of knowledge and application of academic skills in science and mathematics.

Attributes of the STEM Classroom

- Active and student centered
- Equipped to support spontaneous questioning as well as planned investigation
- A center for innovation and invention
- Classroom, laboratory and engineering lab are physically one
- Supportive of teaching in multiple modalities
- Serves students with a variety of learning styles and disabilities
- Integrates real-world situations or problems

Attributes of STEM-Educated Students

- *Problem solvers*—define questions and problems, design investigations to gather data, collect and organize data, draw conclusions, and apply understandings to new and novel situations.
- *Innovators*—creatively use science, mathematics, and technology concepts and principles by applying them to the engineering design process.
- *Inventors*—recognize the needs of the world and creatively design, test, redesign, and implement solutions.
- *Self-reliant*—use initiative and self-motivation to set agendas, develop and gain self-confidence, and work within specified time frames.
- *Logical thinkers*—apply rational and logical thought processes of science, mathematics, and engineering design to innovations and inventions.
- *Technologically literate*—explain the nature of technology, develop the skills needed, and apply technology appropriately.

Although the goal of STEM education is for all four disciplines to be taught in some interdisciplinary manner, in many cases a schism exists between practice and theory. The progress toward a truly integrated interdisciplinary STEM curriculum has been slow to make a foothold as a common occurrence within classrooms (Capraro et al. 2015). Part of this problem may be in how educators define STEM beyond the individual definitions. If a synergy is to be created that goes beyond the individual components of STEM and toward a vision of interdisciplinary STEM, how do we get there?

References

Capraro, M. M., R. M. Capraro, S. Metoyer, S. Nite, and C. A. Peterson. 2015. Promising practices in STEM teaching and learning: A meta synthesis. Paper presented at the STEM Collaborative for Teacher Professional Development, Texas A&M University, College Station, TX.

Honey, M., G. Pearson, and H. Schweingruber. 2014. *STEM integration in K–12 education: Status, prospects, and an agenda for research.* Washington, DC: National Academies Press.

Morrison, J. 2006. TIES STEM Education Monograph Series: Attributes of STEM education— The student, the academy, the classroom. Teaching Institute for Essential Science (TIES). *www. partnersforpubliced.org/uploadedFiles/TeachingandLearning/Career_and_Technical_Education/ Attributes%20of%20STEM%20Education%20with%20Cover%202%20.pdf.*

National Academy of Engineering and National Research Council (NRC). 2002. *Technically speaking: Why all Americans need to know more about technology.* Washington, DC: National Academies Press.

National Academy of Engineering and National Research Council (NRC). 2014. *STEM integration in K–12 education: Status, prospects, and an agenda for research.* Washington, DC: National Academies Press.

Teach Engineering. 2015. Engineering design process. *www.teachengineering.org/engrdesignprocess.php.*

3 | It's About the Question: Matching the *How* to the *Why*

STEM remains the elephant in the room in terms of a new vision to meet the technological challenges of the 21st century. Further iterations and additions of more disciplines (e.g., STEAM, STREAM, STREAMSS) to the STEM cadre has further complicated how educators define and put STEM into practice. The question remains: what does STEM, or its many derivatives, look like in practice? Are there some nonnegotiables to any STEM lesson? Even when the possibility for full integration exists, what should it look like in terms of a lesson? An *Education Week* article (Jolly 2014) identified six characteristics of a great STEM lesson:

1. STEM lessons focus on real-world issues and problems.

2. STEM lessons are guided by the engineering design process.

3. STEM lessons immerse students in hands-on inquiry and open-ended exploration.

4. STEM lessons involve students in productive teamwork.

5. STEM lessons apply rigorous math and science content your students are learning.

6. STEM lessons allow for multiple right answers and reframe failure as a necessary part of learning.

Most STEM lessons found in the public domain do not exhibit all six of these characteristics, and many may not exhibit *any* of these characteristics. But are these six characteristics enough to ensure student engagement and conceptual learning? How these characteristics are introduced in the lesson is critical, and StEMT*ifying* lessons can ensure that the process is in place and the lesson is not fragmented.

Many engineering projects have been used over the years, from building bridges to launching rockets. How many of these projects were more about learning how to use or share glue rather than engineering or science? In many cases, students build a tower or bridge and that is it. Because of the extreme length of time needed to build the first version, the time needed for redesigning and rebuilding the first failed version is a luxury most science teachers do not have. It is certain that the engineering design component of this lesson would be missed as a result, but what about science?

We have seen many examples illustrating engineering in isolation as a common occurrence under the guise of STEM. Engineering design challenges in isolation gives rise to a lack of purpose in terms of science. Most students are quite capable of gluing craft sticks (or whatever the material that is used to build the tower or bridge) together. They can make the tower or bridge as long or tall as necessary. They may even get lucky and choose a strong design based on bridges or towers they have seen. More typical is only testing the strength of the glue joint and not the actual bridge design. So what is learned? Again, just because we can include most or all of the listed criteria, it does not ensure a quality STEM lesson. The reason most of these lessons fail our litmus test of what is a quality STEM lesson is that the *How* does not match the *Why*. Unfortunately, in far too many cases, there is no *Why*, let alone a *Why* matched to a *How*.

To further the point, let us reflect on a very popular engineering design challenge that most teachers of science have seen: the Marshmallow Challenge. The Marshmallow Challenge became extremely popular after the release of the popular *TED Talk* by Tom Wujec (2010). The challenge is to build the tallest possible structure using 20 sticks of spaghetti, one yard of tape, one yard of string and one marshmallow. Teams are given a time limit—typically around 18 minutes—to design and build a tower using the provided materials.

Is the Marshmallow Challenge a STEM lesson? It is marvelous for team building, practicing problem solving, and getting groups to learn to work together based on their combined skills. However, these types of challenges have limited, if any, purpose in teaching science concepts. It is our belief that the purpose, or the *Why*, is the teaching of the science concepts. Within a science classroom, it should be the science that is the driving force of instruction, whether as a siloed or integrated approach. We believe that the integration of math, technology and engineering are to be used to support the learning of a science concept. If the *Why* is about targeting the concepts of science, what about the *How*?

What the science education community knows about learning about science is that it must be explicitly taught and students must be allowed to reflect on what they have learned. In other words, hands-on teaching is not enough; what is required is a hands-on and minds-on approach (Bybee et al. 2006). In a later chapter we discuss the importance of instructional models to target student's preconceptions or misconceptions.

It is critical that we plan explicitly for outcomes rather than hope for them as a mere artifact of instruction. As far back as Herbart (1901) it was realized that any new knowledge needs to be connected to prior knowledge. Good pedagogical practice also allows students to make connections with other experience, more recently referred to as crosscutting concepts (*A Framework for K–12 Science Education;* NRC 2012). Herbart (1901) goes on to explain that students must be allowed to apply or demonstrate their knowledge in some meaningful and relevant fashion, seemingly making an early case for problem-based or project-based learning. Over 100 years ago, the idea that teaching should be intentional and explicit was known, as well as the use of relevant themes needed to help students make connections between abstract concepts and the world around them.

As learning occurs against a backdrop of previous conceptions, students must first become aware of their fundamental assumptions (Posner et al.1982). Posner et al. (1982) also suggest that the following conditions need to be met before conceptual change has a possibility of taking place:

1. *There must be dissatisfaction with existing conceptions.*

2. *A new conception must be intelligible.*

3. *A new conception must appear initially plausible.*

4. *A new concept should suggest the possibility of a fruitful research program.*

In Chapter 5, the use of the 5E Instructional Model will be discussed as a means to make learning intentional. Modern education has given us sets of knowledge that students are expected to master, typically in the form of standards. These standards are what educators should target when planning lessons. As we know the outcomes that are expected from the educational experience of the students, it seems obvious that the lessons created are intentionally explicit to help students reach the goal of a deep understanding of the construct being surveyed and to help students make connections to real-world situations. Dewey expressed the need for the lesson to begin the learning by causing students to perplexity, confusion, and doubt about a certain phenomenon or situation (Dewey 1910). Dewey's instructional model (1910) requires the teacher to present an experience where the students feel thwarted and sense a problem. The more recent 5E Instructional Model (to be discussed later) requires students to mentally focus on an object, problem, situation, or event. The activities of this phase make connections to past experiences and expose students' misconceptions (BSCS and IBM 1989).

Focusing a lesson on the appropriate question or questions is critical to constructing effective STEM lessons. What is it that we want our students to take away from the lesson? A STEM lesson is not different from other lessons. There is a plethora of research on how students learn that cannot and should not be abandoned. If the goal is the application, extension, and additional support for learning through the STEM process, the *Why* (we are doing the lesson) must be actively connected to the *How* (we facilitate the student learning).

Although the purposes of the *Why* needs to be matched with the *How* in terms of fully and clearly expressing or demonstrating the targeted science concepts, what are some other variables? You might have noticed that in the list of conditions needed to make a good STEM lesson, integration was absent from the list, but why? We come back to the important question with which we began this dialogue: what is STEM? Is it about integration? How is integration related to a good STEM lesson?

References

BSCS and IBM. 1989. *New designs for elementary science and health: A cooperative project between Biological Sciences Curriculum Study (BSCS) and International Business Machines (IBM).* Dubuque, IA: Kendall Hunt.

Bybee, R. W., J. A. Taylor, A. Gardner, P. VanScotter, J. Carlson Powell, A. Westbrook, and N. Landes. 2006. *The BSCS 5E Instructional Model: Origins and effectiveness.* Colorado Springs, CO: Biological Sciences Curriculum Study.

Dewey, J. (1910) 1971. *How we think.* Chicago: Henry Regnery Company.

Herbart, J. 1901. *Outlines of educational doctrine*, ed. A. Lange, trans. C. DeGarmo. New York: Macmillan.

Jolly, A. 2014. Six characteristics of a great STEM lesson. *Education Week. www.edweek.org/tm/articles/2014/06/17/ctq_jolly_stem.html.*

National Research Council (NRC). 2012. *A framework for K–12 science education: Practices, crosscutting concepts, and core ideas.* Washington, DC: National Academies Press.

Posner, G. J., K. A. Strike, P. W. Hewson, and W. A. Gertzog. 1982. Accommodation of a scientific conception: Toward a theory of conceptual change. *Science Education* 66 (2): 211–227.

Wujec, T. 2010. Build a tower, build a team. TED. *www.ted.com/talks/tom_wujec_build_a_tower?utm_source=tedcomshare&utm_medium=referral&utm_campaign=tedspread.*

CHAPTER
4 | Is Integration Enough?

Is STEM just integration? Even the acronym *STEM* provides few clues as to how to integrate these disciplines. So what makes a lesson a STEM lesson? Is STEM the inclusion of a more focused math component within a science lesson, or is that merely good science instruction? If you used graphing calculators, probeware, or computers, would that bring it nearer to the STEM goal? Again, this is good science instruction, not necessarily STEM. Does incorporating an engineering challenge define a STEM lesson, or does this activity support the original science concept? One could make the point that even if technology is integrated into the lesson as a separate concept from the original science lesson, it is STEM; however, STEM is not just integration of technology (National Academy of Engineering 2014). One may argue that any integration of technology can turn a discipline-specific lesson into STEM. We share the same stance as the National Academy of Science and NRC (2014) that STEM must be more than just mere integration.

Going back to the original definition of science, science is defined as an investigation of the natural world. So what are the questions a scientist asks compared to the questions an engineer asks? A scientist's quest is to understand the natural world, but an engineer's task is to improve on the natural world for the benefit of humans. For example, a scientist may study a new compound to see how it interacts with other atoms or molecules. An engineer looks at how this new compound can be used. It is this difference between science and its application that we feel is the crux of STEM. How can we use engineering and its resulting technology to support the learning of science concepts? Technology within STEM lessons as the outcome and not the inclusion within instruction may be the answer.

We argue that merely integrating technology cannot be the intention of STEM education. It would seem difficult to understand how the act of integration of technology into a science lesson gets us any closer to the goal of the democratic vision (Driver et al. 1996), so commonly referred to in educational reforms as an argument for students becoming problem solvers or good consumers of information. It would seem obvious that to become a problem solver, a student would need experience in solving problems. Merely integrating some or all disciplines of STEM into a lesson would seem insufficient to meet these lofty goals. Even *A Framework for K–12 Science Education* (*Framework*); NRC 2012) states,

> *Engineering and technology are featured alongside the physical sciences, life sciences, and earth and space sciences for two critical reasons: to reflect the importance of understanding the human-built world and to recognize the value of better integrating the teaching and learning of science, engineering, and technology.*

Figure 4.1. The BSCS 5Es as a cycle of learning

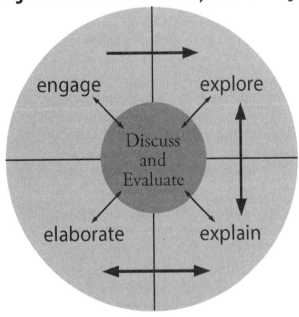

Source: Morgan and Ansberry (2017), as adapted from Barman (1997). *The learning cycle revised: A modification of an effective teaching model.* Arlington, VA: Council for Elementary Science International.

Many in science education are familiar with the 5E Instructional Model (Bybee et al. 2006; see Figure 4.1).

But can the engineering component of the lesson be used as an activator and a reason to learn about the science? Certainly, building and watching a bridge break may be a wonderful hook for students and a wonderful platform for teachers to start asking questions about the underlying scientific principles. Is that STEM? Have students actually learned anything about scientific principles while building the initial bridge? Has the activator brought students misconceptions to the forefront? Would using a bridge stress simulation to bring in the technology component be any more effective? These are the types of questions we predominantly hear about the current state of STEM education. But is STEM just integration? Even the acronym *STEM* provides few clues as to how to integrate these disciplines. Is that STEM? Have students actually learned anything about scientific principles while building the initial bridge?

According to Thomas Hughes' book *Human-Built World* (2004), engineering is in place because of the needed understanding of the human-built world. In those terms, the purpose of engineering in STEM would seem to require the construction (either fabricated or conceptual modeling) of technological ideas. The *Framework* (NRC 2012) also includes the necessity for the inclusion of technology alongside the disciplines of science and engineering. So what does integration mean for the learning of science? Where does this leave us in terms of developing effective STEM lessons?

Rearranging the *STEM* Acronym to Facilitate Effective Instruction

Science (S) is a body of knowledge that seeks to describe and understand the natural world and its physical properties. Scientific knowledge can be used to make predictions, and science uses a process to generate knowledge. *Engineering* (E) is design under constraint and seeks solutions for societal problems and needs. Engineering aims to produce the best solution (to the question posed by science) given resources and constraints, and engineering uses a process—the engineering design process—to produce solutions and technologies. *Mathematics* (M) is the science of quantity that seeks out patterns and uses abstraction and logical reasoning to explore, think through an issue, and reason logically to solve routine as well as nonroutine problems. Engineers must use mathematics to design new technologies. *Technology* (T) is the body of knowledge, systems, processes, and artifacts that result from engineering. Almost everything made by humans to solve a need or fulfill a desire is a technology, including pencils, shoes, cell phones, and processes to treat water. Science poses the question, and through the use of engineering design and mathematics, a new technology is developed to solve the question posed by science. Science (infused with instructional technology tools[t]), engineering, mathematics, and technology (StEMT) is a *process*.

A traditional STEM lesson may use the following question to guide instruction: *How does the amount of heat energy affect water?* An example of a StEMT*ified* question may be, "How can phase change and the states of matter be used to help provide clean water to people in Third World countries?"

The original question could be used during the first section of the lesson. The second, problem-based question is introduced after the concept has been mastered and students are ready to apply their knowledge to a real-world problem. This approach places StEMT*ification* in the elaborate section of a 5E lesson, where engineering design and problem solving are used to make relevant meaning of the science concept. To be clear, we make no claim that engineering is the only activity that can live within an elaborate section of a 5E lesson. If including an engineering component with a lesson makes logical sense, the elaborate section may be a very appropriate place to include it. In some cases, engineering may not be what is needed for students to develop the deep understanding of a concept. In these cases, you may decide to include other extension activities. For example, you may choose to include a research component where students delve into situational science. In the lessons in the following chapters, we have included a small section that includes other common types of activities we have seen over the years appearing in lesson plans. We feel that the lessons we have chosen are ones that can benefit from an engineering component.

References

Bybee, R. W., J. A. Taylor, A. Gardner, P. VanScotter, J. Carlson Powell, A. Westbrook, and N. Landes. 2006. *The BSCS 5E Instructional Model: Origins and effectiveness.* Colorado Springs, CO: Biological Sciences Curriculum Study.

Driver, R., J. Leach, R. Millar, and P. Scott. 1996. *Young people's images of science.* Bristol, PA: Open University Press.

Hughes, T. P. 2004. *Human-built world: How to think about technology and culture.* Chicago: University of Chicago Press.

Morgan, E., and K. Ansberry. 2017. *Picture-perfect STEM lessons, K–2: Using children's books to inspire STEM learning.* Arlington, VA: NSTA Press.

National Academy of Engineering and National Research Council (NRC). 2014. *STEM integration in K–12 education: Status, prospects, and an agenda for research.* Washington, DC: National Academies Press.

National Research Council (NRC). 2012. *A framework for K–12 science education: Practices, crosscutting concepts, and core ideas.* Washington, DC: National Academies Press.

CHAPTER

5 | Teaching StEMT Through Inquiry and the 5E Instructional Model

The 5E Instructional Model (Bybee 1993, 2015) is designed to incorporate all aspects of inquiry learning environments by *engaging* students and allowing students to *explore* the concepts being introduced, discover *explanations* for the concepts they are learning, and *elaborate* on what they have learned by applying their knowledge to new situations. According to constructivist learning theory, knowledge is constructed as students integrate new information with their pre-existing knowledge base (Bodner 1986).

5E Instructional Model

I don't know what's the matter with people: they don't learn by understanding, they learn by some other way—by rote or something. Their knowledge is so fragile!

—Richard Feynman, Nobel Award–winning physicist

The quote above describes knowledge and the way it is gained by many people. To completely understand what is meant by this statement, the sentences must be deconstructed. What is meant by learning through *rote*, another way of saying that humans have a tendency to memorize bits of information? The term *fragile* would seem to apply a sense of tentativeness, but why is it fragile?

From a learning perspective, situations like these can occur when a person's knowledge is shallow. For instance, a person may know that light gets reflected off of some items and not others. If that same person is asked how his or her reflection might look in flat or curved mirrors, that person may have no idea. Worse, the ideas about what his or her reflection might look like are filled with misconceptions pertaining to the principles of light. How does one gain knowledge beyond the rote or superficial?

In terms of science learning, inquiry has been a preferred method of instruction for decades. It was at the forefront of the *National Science Education Standards* (NRC 1996) and *Inquiry and the National Science Education Standards* (NSES 2000).

> *Scientific inquiry refers to the diverse ways in which scientists study the natural world and propose explanations based on the evidence derived from their work. Inquiry also refers to the activities of students in which they develop knowledge and understanding of scientific ideas, as well as an understanding of how scientists study the natural world.* (NRC 1996, p. 23)

With decades of research behind it, educators realize the power of inquiry. Through inquiry, students are able to interact with a phenomenon. They observe and discuss these concepts with their peers, providing them opportunities to experience and explain this phenomenon from many different perspectives. Currently, one of the favored methods for teaching science is through the 5E Instructional Model developed by the Biological Sciences Curriculum Study (BSCS) in 1987 (Bybee et al. 2006, Bybee 2015).

Figure 5.1. The 5E Instructional Model developed by BSCS

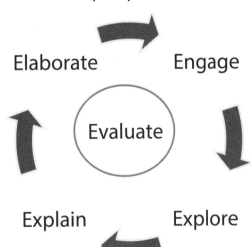

Source: Bybee et al. (2006).

It is not surprising that research supports what science educators have known for many years—that a research-based instructional model aligned to what is known about how students learn is critical. This instructional model also must be used consistently to obtain the types of outcomes that are desired (Bransford, Brown, and Cocking 2000).

For more than three decades, the 5E Instructional Model has been used extensively in the development of new curriculum materials as well as professional development opportunities (Bybee et al. 2006, Bybee 2015). The model, commonly referred to as the BSCS 5E Instructional Model (Figure 5.1), or the 5Es, consists of five steps.

Each step has a specific purpose in the learning process, all designed to help students move beyond their misconceptions. Misconceptions noted in the lessons are adapted from the American Association for the Advancement of Science (AAAS Project 2061 n.d.) The steps for 5E listed in Table 5.1 are designed to provide opportunities for the learner to develop a better understanding of scientific and

Table 5.1. Summary of the BSCS 5E Instructional Model

Phase	Summary
Engage	The teacher or a curriculum task captures students' attention by accessing their prior knowledge to help them become engaged in a new concept through the use of short activities that promote curiosity and elicit prior knowledge. The activity should make connections between past and present learning experiences, expose prior conceptions (or misconceptions), and organize students' thinking toward the learning outcomes of current activities. This experience must have a meaningful context, and the experience should generate student questions leading to teachable moments.
Explore	Exploration experiences provide students with a common base of activities within which current concepts (i.e., misconceptions), processes, and skills are identified and conceptual change is facilitated. Learners may complete lab activities that help them use prior knowledge to generate new ideas, explore questions and possibilities, and design and conduct a preliminary investigation. This is the time for students to "resolve the mental disequilibrium" of the Engage experience by trying to answer their engaging questions.
Explain	The explanation phase focuses students' attention on a particular aspect of their engagement and exploration experiences and provides opportunities to demonstrate their conceptual understanding, process skills, or behaviors. This phase also provides opportunities for teachers to directly introduce a concept, process, or skill. Learners explain their understanding of the concept. An explanation from the teacher or the curriculum may guide them toward a deeper understanding, which is a critical part of this phase. Most important, this phase allows students to express their explanations and allows the teacher to use teachable moments.
Elaborate	Teachers challenge and extend students' conceptual understanding and skills. Through related but new experiences, the students develop deeper and broader understanding, more information, and adequate skills. Students apply their understanding of the concept by conducting additional activities and connecting to the explanation.
Evaluate	The evaluation phase encourages students to assess their understanding and abilities and provides opportunities for teachers to evaluate student progress toward achieving the educational objectives.

Source: BSCS-Biological Sciences Curriculum Study (Bybee 2015)

technological knowledge, attitudes, and skills. For the teacher, the model should be used to organize and sequence programs, units, and lessons.

Given what we know about how instruction should look inside the classroom and the research-based effectiveness of the 5E Instructional Model, we argue that the model should be used in all instruction that includes the learning of science concepts. This argument extends to the use of engineering projects within any STEM program. The question then remains as to how to merge an instructional model designed to break misconceptions about science concepts and the introduction of the popular engineering processes that are now included in STEM. So where does StEMT fit into 5E?

Although a few things need to happen before the insertion of StEMT into 5E, the basic premise is that engineering lives in the Elaborate portion of a 5E lesson plan. Engineering is not a separate entity and is included to support the learning of science and not just extend it. Figure 5.2 illustrates the proposed instruction of the StEMT conceptual framework with the instructional model of 5E.

Figure 5.2. StEMT conceptual framework and 5E Instructional Model

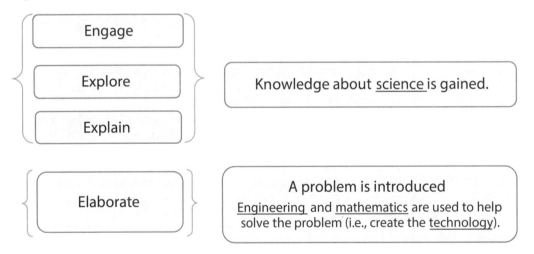

This is only a guide for integrating STEM activities through the process of StEMT. Every science lesson does not need to integrate all aspects of STEM. In some cases, an introductory engineering activity could be used to engage students and activate prior knowledge. The guiding question for the lesson might be based on the problem or it could be introduced as a secondary question at the beginning of the Elaborate section. There is not one right way to use StEMT. The critical difference in implementing StEMT is making sure that the questions drive the lessons.

Inquiry

Inquiry and the National Science Education Standards : A Guide for Teaching and Learning (NRC 2000) was written as a practical guide for teachers, professional developers, administrators, and others who wanted to increase the emphasis on inquiry in science classrooms. Through examples and discussion, *Inquiry and the National Science Education Standards* showed educators how students and teachers can use inquiry to learn how to do science, learn about the nature of science, and learn science content. Although the integration of STEM into science lessons was not the purpose of *Inquiry and the National Science Education Standards*, the discussion about the essential features of inquiry and the possible variations is fundamental to understanding the many facets of inquiry.

Students who use inquiry to learn science engage in many of the same activities and thinking processes as scientists who are seeking to expand human knowledge of the natural world (NRC 1996). In *Inquiry and the National Science Education Standards* (NRC 2000, p. 19), the fundamental abilities necessary for students in grades 5–8 to do scientific inquiry are identified:

- Identify questions that can be answered through scientific investigations.

- Design and conduct a scientific investigation.

- Use appropriate tools and techniques to gather, analyze, and interpret data.

- Develop descriptions, explanations, predictions, and models using evidence.

- Think critically and logically to make the relationships between evidence and explanations.

- Recognize and analyze alternative explanations and predictions.

- Communicate scientific procedures and explanations.

- Use mathematics in all aspects of scientific inquiry.

These fundamental abilities for inquiry can be recognized within the *Common Core State Standards for Mathematics* and the *Framework for K–12 Science Education* (NRC 2012) science and engineering practices. However, a lingering question posed in Chapter 2 of *Inquiry and the National Science Education Standards* was "What is teaching through inquiry, and when and how should it be done?"

> *In the classroom, a question robust and fruitful enough to drive an inquiry generates a "need to know" in students, stimulating additional questions of "how" and "why" a phenomenon occurs. Fruitful inquiries evolve from questions that are meaningful and relevant to students, and the knowledge and procedures students use to answer the ques-*

tions must be accessible and manageable, as well as appropriate to the students' developmental level (NRC 2000, pp. 24–25).

First, as you might imagine, you start with a good science lesson. From there, you decide on what types of problems might be able to be solved through gaining the deeper understanding of science within the original science lesson. The next steps involve some creativity. You will need to decide on the parameters of what students will be able to do within their science classroom that can provide opportunities to solve the problem using their newly attained science knowledge. Once you have decided on the lesson and the engineering design you want students to do, you will need to construct the guiding question that will drive the lesson. Most lessons already have a guiding question that frames the lesson. This question may or may not be adequate for StEMT*ifying* your lesson. We have found that in the vast majority of cases, it is not. That is not a criticism of the original lesson. Remember, we are now stretching the lesson from inquiry (learning about a phenomenon) to problem-solving a solution using the knowledge gained from the inquiry.

Inquiry and Differentiation

By its very nature, inquiry is problem-based learning, and as such the process of inquiry and StEMT allows for varied learning strengths requiring the use of a range of resources and provides opportunity for student choice with teacher guidance. Group work also lends itself to differentiation by learner interest, ability levels, and assigned roles (Tomlinson 1999, 2001). The 5E Instructional Model, inquiry, and the StEMT process allow for a tiered approach also; teachers can focus on essential understandings (disciplinary core ideas) and skills at different levels of complexity, abstractness, and open-endedness (Tomlinson 1999). Many students show what they know better in a product than a written test, and high-quality assignments (products resulting from the StEMT process) are excellent ways of assessing student knowledge, understanding, and skill (Tomlinson 2001).

A good product is not just something students do for enjoyment at the end of a unit. It must cause students to think about, apply, and even expand on all the key understandings [disciplinary core ideas] and skills [practices] of the learning span it represents. Because the product assignment should stretch students in application of understanding and skill as well as the pursuit of quality, a teacher needs to determine ways in which she can assist the student in reaching a new level of possibility as the product assignment progresses. This sort of scaffolding allows students to find success at the end of hard work rather than overdoses of confusion and ambiguity. (Tomlinson 2001, pp. 85–86)

Teaching StEMT Through Inquiry and the 5E Instructional Model

Teachers and students can differentiate product assignments based on student readiness, interest, and learning profile, but the structure needs to focus and guide students while allowing the freedom for innovation, thought, and creativity (Tomlinson 1999, 2001). The teacher knows the students and their differentiation needs, and although most lessons average five 45-minute class periods, students should be given additional time if necessary. If needed, students can be provided with writing templates for summative writing activities. However, students should first try writing on their own before being given the template.

It only makes sense that if we want students to become problem solvers, they must be provided with the opportunities (NRC 2012). So how do we change the guiding question from one that frames inquiry to one that portrays a problem? Once you know the type of engineering project to be completed, it is only a matter of working backward to construct the appropriate question. For example, the following lesson was originally written to help third-grade students understand phase change and the states of matter. The engineering project we chose to use was water purification. A brief StEMT*ified* 5E lesson example is shown in Table 5.2.

Table 5.2. StEMT*ified* 5E Lesson

Guiding question (Original): How does the amount of heat energy affect water?
StEMTified question: How can phase change and the states of matter be used to help provide clean water to people in Third World countries?
ENGAGE
Ask the students to make a list of places they can find water. Do NOT correct misconceptions at this time. Read aloud a short article about water pollution issues in Third World countries.
EXPLORE
Step 1: Give each student a cup with one ice cube and ask them to put the pieces of ice on their tongues but not to chew them.
Step 2: Give each student a baggie with an ice cube and ask them to melt the ice cube without taking it out of the bag.
Step 3: Place an ice cube on a hot plate or in a microwave. Record observations for each activity.

Continued

Table 5.2 (*continued*)

EXPLAIN
On a paved area, have the students stand as close to each other as they can while you draw a chalk circle around them. Students move without getting out of the circle (solid). Draw another circle outside of the original circle. Students step away from each other so there is at least an arm's length between them and they can no longer touch one another (liquid). Students spread out in the entire area of a large given space and move around the space (gas).
ELABORATE (It is here that the major work of StEMT*ifying* the lesson will happen.)
Original activity for this lesson: Ask students to estimate how long it would take the ice cube to melt at room temperature, in the sunlight, in a cooler, or in a refrigerator. Invite other suggestions and explore as time allows.
StEMT*ify* and address the six characteristics of a STEM lesson.
Restate the original StEMT*ified* guiding question: How can phase change and the states of matter be used to help provide clean water to people in Third World countries?
Present the students with the engineering design challenge: The goal is to use their knowledge of how water changes from its different states to build an efficient working model of a water purification system. Students will be provided with a set of materials. Their energy source will be the Sun. A heat lamp can be substituted in cooler climates or on cloudy days. It is up to the students to figure out the appropriate design to turn the liquid water to water vapor. They will then need to collect that water vapor and allow it to condense so that it can be recaptured. Extension question: Is your design the most efficient system possible? Demonstrate how the design is an effective way of producing clean water. Redesign to make it more effective.
EVALUATE
Formative assessment occurs continuously through the lesson and engineering design challenge.
Summative assessment: Write About It! Claims, Evidence, Reasoning (CER): Was the design efficient (water was purified)? What evidence did you collect to determine that the water purification design is efficient? How can this design help people in countries where clean water is not readily available?

Source: Adapted with permission from the Biological Sciences Curriculum Study (BSCS 2015).

What you might have noticed was the absence of technology in this lesson. It certainly could easily be added with the use of probeware or the use of laptops to collect and write down data. Adding technology is a great option and should be considered for every lesson, but it is not critical, as the technology we promote is the solution to the problem. This is where technology is created rather than just used. It is here that we move students from knowledge gatherers to problem solvers.

As you can see from the previous example, StEMT*ifying* a lesson is really easy. It can be as simple as changing or adding a question and inserting an engineering design challenge into the Elaborate portion of a 5E lesson. Again, our belief is that the engineering portion is added to support the learning of the science concept. It is not, nor do we believe that it ever should be, placed in a science lesson for any other reason.

References

AAAS Project 2061. n.d. Pilot and field test data collected between 2006 and 2010. Unpublished raw data.

Biological Sciences Curriculum Study (BSCS). 2015. BSCS 5E Instructional Model. *http://bscs.org/bscs-5e-instructional-model.*

Bodner, G. M. 1986. Constructivism: A theory of knowledge. *Journal of Chemical Education* 63 (10): 873–878.

Bransford, J., A. Brown, and R. Cocking, eds. 2000. *How people learn: Brain, mind, experience, and school.* Washington, DC: National Academies Press.

Bybee, R. 1993. An instructional model for science education. In *Developing biological literacy.* Colorado Springs, CO: Biological Sciences Curriculum Study.

Bybee, R. W. 1993. *Reforming science education: Social perspectives and personal reflections.* New York: Teachers College Press.

Bybee, R. W. 2009. *The BSCS 5E Instructional Model and 21st century skills.* Washington, DC: National Academies Press.

Bybee, R. W. 2010. What is STEM education? *Science* 329 (5995): 996.

Bybee, R. W. 2015. *The BSCS 5E Instructional Model: Creating teachable moments.* Arlington, VA: NSTA Press.

Bybee, R. W., J. A. Taylor, A. Gardner, P. VanScotter, J. Carlson Powell, A. Westbrook, and N. Landes. 2006. *The BSCS 5E Instructional Model: Origins and effectiveness.* Colorado Springs, CO: Biological Sciences Curriculum Study.

National Research Council (NRC). 1996. *National science education standards.* Washington, DC: National Academies Press.

National Research Council (NRC). 2000. *Inquiry and the national science education standards: A guide for teaching and learning.* Washington, DC: National Academies Press.

National Research Council (NRC). 2012. *A framework for K–12 science education: Practices, crosscutting concepts, and core ideas.* Washington, DC: National Academies Press.

Tomlinson, C. A. 1999. *The differentiated classroom: Responding to the needs of all learners.* Alexandria, VA: ASCD.

Tomlinson, C. A. 2001. *How to differentiate instruction in mixed ability classrooms.* 2nd ed. Alexandria, VA: ASCD.

CHAPTER
6 | Correlation to *A Framework for K–12 Science Education*

The 2012 publication *A Framework for K–12 Science Education* (*Framework;* NRC 2012) outlines the three dimensions that provide students with a context for the content of science, demonstrate how science knowledge is acquired and understood, and show how the sciences are connected through concepts that have universal meaning across the disciplines:

- Scientific and engineering practices;

- Crosscutting concepts that unify the study of science and engineering through their common application across fields; and

- Core ideas in four disciplinary areas: physical sciences; life sciences; Earth and space sciences; and engineering, technology, and the applications of science.

To support students' meaningful learning in science and engineering, all three dimensions need to be integrated into standards, curriculum, instruction, and assessment. Engineering and technology are featured alongside the natural sciences (physical sciences, life sciences, and earth and space sciences) for two critical reasons: to reflect the importance of understanding the human-built world and to recognize the value of better integrating the teaching and learning of science, engineering, and technology (*Framework;* NRC 2012).

The scientific and engineering practices identified in the *Framework* are the major practices that scientists employ as they investigate and build models and theories about the world and a key set of engineering practices that engineers use as they design and build systems. The practices specify what is meant by inquiry in science and the range of cognitive, social, and physical practices that it requires (NRC 2012). As in all inquiry-based approaches to science teaching, the expectation is that students will engage in the practices. Students cannot comprehend scientific practices without directly experiencing those practices.

Science and Engineering Practices (Framework, Dimension 1)

1. Asking questions (science) and defining problems (engineering)
2. Developing and using models
3. Planning and carrying out investigations
4. Analyzing and interpreting data
5. Using mathematics, information and computer technology, and computational thinking
6. Constructing explanations (science) and designing solutions (engineering)
7. Engaging in argument from evidence
8. Obtaining, evaluating, and communicating information

Standards for Mathematical Practice (Common Core State Standards)

1. Make sense of problems and persevere in solving them
2. Reason abstractly and quantitatively
3. Construct viable arguments and critique the reasoning of others
4. Model using mathematics
5. Use appropriate tools strategically
6. Attend to precision
7. Look for and make sense of structure
8. Look for and express regularity in repeated reasoning

The practices, whether engineering, science, or mathematics, are the process standards of problem solving, reasoning and evidence, communication, modeling, and applications.

Evaluating the Fidelity of 5E Inquiry and StEMT

The *Common Core State Standards for Mathematics* (*CCSS Mathematics*; NGAC and CCSSO 2010) are an integral part of learning and doing mathematics and need to be taught with the same intention and attention as mathematical content. The science and engineering practices (Dimension 1: Practices) identified in the *Framework* (NRC 2012) are derived from those that scientists and engineers actually engage in as part of their work. The mathematics, science, and engineering practices are not intended to be taught as stand-alone lessons. The practices are an integral part of learning and doing in all content areas and need to be taught with the same intention and attention as the content standards. Opportunities for students to immerse themselves in these practices and to explore why they are central to mathematics, science, and engineering are critical to appreciating the skill of the expert and the nature of his or her enterprise. Mayes and Koballa (2012) illustrated one way the practices can be aligned (see Table 6.1, p. 30).

School districts across the nation use evaluation systems for instructional personnel that focus on specific domains of teacher and student behaviors. For the purposes of this discussion, the *Teacher Evaluation Model* (Marzano 2007) and *The Framework for Teaching* (Danielson, 2013) will be referenced in the teacher behaviors for effective instructional delivery and facilitation, specifically *Domain 1: Classroom Strategies and Behaviors* (Marzano) and *Domain 3: Engaging Students in Learning* (Danielson). However, the process of administrator classroom walkthroughs and what they should "look for" in a 5E instructional delivery model that integrates the StEMT process are applicable to any evaluation model. The *Look Fors* for student behaviors (Table 6.2, p. 31) and teacher behaviors (Table 6.3, p. 31) that reflect sound habits of mind and support the implementation of the 5E instructional model and StEMT process are adapted from the *Framework* Science and Engineering Practices (SEP) and *Common Core State Standards for Mathematics* (MP).

The graphic in Figure 6.1 (p. 32) shows the relationships and convergences (commonalities) found in the *CCSS Mathematics* (designated as MP), the English Language Arts Proficiency Development Framework (ELA practices designated as EP), and the science and engineering practices identified in the *Framework* (designated as SEP).

Notice that the center where the three converge require the use of evidence to assert a claim through reasoning:

- EP 1: Demonstrate independence.

- MP 3: Construct viable arguments and critique reasoning of others.

- EP 3: Respond to the varying demands of audience, task, purpose, and discipline.

- SEP 7: Engage in argument from evidence.

The practices, crosscutting concepts, and disciplinary core ideas (NRC, 2012) have application across all domains of science and mathematics:

Table 6.1. Alignment of science, engineering, and mathematical practices

Science and Engineering Practices *(Framework, Dimension 1)*	Mathematical Practices *(CCSS)*
1. Asking questions and defining problems 3. Planning and carrying out investigations	Make sense of problems and persevere in solving them.
2. Developing and using models 3. Planning and carrying out investigations 5. Using mathematics and computational thinking	Reason abstractly and quantitatively.
5. Using mathematics and computational thinking 6. Constructing explanations and designing solutions 7. Engaging in argument from evidence 8. Obtaining, evaluating, and communicating information	Construct viable arguments and critique the reasoning of others.
2. Developing and using models 3. Planning and carrying out investigations	Model with mathematics.
2. Developing and using models 3. Planning and carrying out investigations 4. Analyzing and interpreting data	Use appropriate tools strategically.
3. Planning and carrying out investigations 8. Obtaining, evaluating, and communicating information	Attend to precision.
4. Analyzing and interpreting data 6. Constructing explanations and designing solutions 7. Engaging in argument from evidence	Look for and make sense of structure.
5. Using mathematics and computational thinking 6. Constructing explanations and designing solutions	Look for and express regularity in repeated reasoning.

Source: Adapted from Mayes and Koballa (2012).

Table 6.2. Student behaviors incorporating science, engineering, and mathematics

Students should
• Ask questions, define problems, and predict solutions/results (SEP.1/MP.1).
• Design, plan and carry out investigations to collect and organize data (e.g. science notebook/journal) (SEP 3/MP 1).
• Develop and use models (SEP 2/MP 4).
• Obtain, evaluate, and communicate information by constructing explanations and designing solutions (SEP 8/MP 3).
• Be actively engaged and work cooperatively in small groups to complete investigations, test solutions to problems, and draw conclusions. Use rational and logical thought processes, and effective communication skills (writing, speaking and listening) (SEP 7, SEP 8/MP 3).
• Analyze and interpret data to draw conclusions and apply understandings to new situations (SEP 4/MP 5).
• Use mathematics, information technology, computer technology, and computational thinking in a creative and logical manner (SEP 5/MP 2).
• Acquire and apply scientific vocabulary *after* exploring a scientific concept (SEP 6/MP 7).

Table 6.3. Teacher behaviors correlated to Marzano (2007) and Danielson (2013) evaluation models

Teachers should
• Integrate real-life applications and subject matter to exemplify how the disciplines co-exist in actual practice (*relate and integrate the subject matter with other disciplines and life experiences*).
• Incorporate experimental design and engineering design processes for all students (*engaging students in learning*).
• Deliver standards-based curriculum using appropriate pedagogy/instructional materials.
• Introduce scientific vocabulary **after** students have had the opportunity to explore a scientific concept (*integrating content area literacy strategies and application of the subject matter*).
• Ask guiding questions to facilitate discussion and active engagement in inquiry, scientific process, and problem solving (*using questioning and discussion techniques*).
• Facilitate questioning and testing solutions to problems. Encourage collaboration so all group members are actively engaged (*communicating with students* and *employ higher-order questioning techniques*).
• Move around the room, guiding cooperative learning groups in formulating solutions and using manipulatives/technology (*establishing a culture for learning, respect, and rapport*).
• Use formative and summative assessments that focus on problem solving and deep understanding rather than memorizing facts.

Figure 6.1. Commonalities among the science, mathematics, and English language arts (ELA) practices

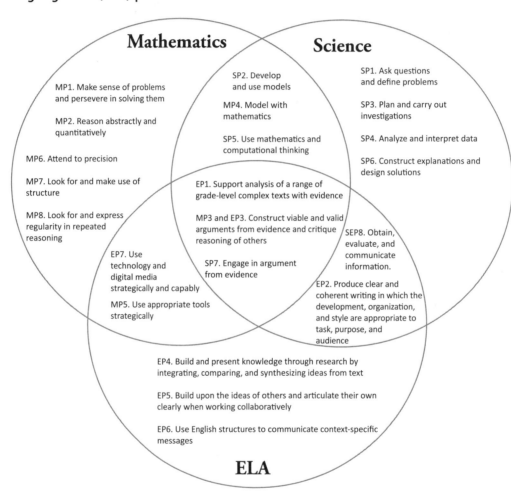

Source: Adapted with permission from Cheuk (2013).

Dimension 1: Science and Engineering Practices (identified in Table 6.1 on p. 30 and Figure 6.1 on p. 32)

Dimension 2: Crosscutting Concepts

- *Patterns*

- *Cause and effect: Mechanism and explanation*

- *Scale, proportion, and quantity*

- *Systems and system models*

- *Energy and matter: Flows, cycles, and conservation*

- *Structure and function*

- *Stability and change*

Dimension 3: Disciplinary Core Ideas (organization of StEMT lessons in the next three chapters)

STEM education should focus on a limited number of disciplinary core ideas and crosscutting concepts, be designed so that students continually build on and revise their knowledge and abilities over multiple years, and support the integration of such knowledge and abilities with the practices needed to engage in scientific inquiry and engineering design (NRC 2012). The *Framework* defines a core idea for K–12 science instruction:

- Have broad importance across multiple sciences or engineering disciplines or be a key organizing principle of a single discipline.

- Provide a key tool for understanding or investigating more complex ideas and solving problems.

- Relate to the interests and life experiences of students or be connected to societal or personal concerns that require scientific or technological knowledge.

- Be teachable and learnable over multiple grades at increasing levels of depth and sophistication—that is, the idea can be made accessible to younger students but is broad enough to sustain continued investigation over the years.

Integration should be intentional and seamless across representations and materials. These multi-day units must also provide explicit support for students as they build knowledge and skill within and across disciplines. All StEMT lessons are correlated to the practices, disciplinary core ideas, and crosscutting concepts in the *Framework*.

Connecting ideas across disciplines is challenging when students have little or no understanding of the relevant ideas in specific disciplines and do not use their disciplinary knowledge in integrated contexts. Students need support to elicit the relevant scientific or mathematical ideas in an engineering or technological design context, to connect those ideas productively, and to reorganize their own ideas in ways that come to reflect normative scientific ideas and practices (NRC 2012).

The next three chapters contain sample lessons that have gone through the process of being StEMT*ified*. The basic premise or activity in the lessons are not unique—many of these lessons have been used by teachers across the nation for many years, in some form or another. Where appropriate, free open educational resources (OERs) are referenced. The important piece on which to focus is the process of embedding the engineering component into more traditional lessons. By some definitions of STEM, the lessons we have started with could be thought of as STEM lessons, but without the engineering component they may lack a way to elicit the real-world discussion and relevance of engineering design for student learning. It is this facet of student understanding that is critical to long-term retention of important concepts.

A quick word of caution before you begin using the following lessons, which come from our own experiences in the classroom. You will need to answer for yourself this question: *Do your students really understand the engineering design process?* If the answer to this question is *no* or even *maybe*, we would urge you to review the process with your students before you embark on these lessons. The good news is that students will come to an understanding quickly.

The StEMT*ified* sections of each lesson the adhere to the following format from TeachEngineering.org:

- *Step 1: Ask—Practice of asking questions (science) and defining problems (engineering)*

Students will ask questions, define problems, and predict solutions/results (SEP 1; MP 1). What is the problem to solve? What needs to be designed and who is it for? What are the project requirements and limitations? What is the goal?

- *Step 2: Research and Design—Practice of planning and carrying out investigations*

Students will be actively engaged and work cooperatively in small groups to complete investigations, test solutions to problems, and draw conclusions. Use rational and logical thought processes, and effective communication skills (writing, speaking and listening) (SEP 7, SEP 8; MP 3). Students collaborate to brainstorm ideas and develop as many solutions as possible.

- *Step 3: Plan—Practice of constructing explanations and designing solutions*

Students will design, plan, and carry out investigations to collect and organize data (SEP 3; MP 1). Students will compare the best ideas and then select one solution and make a plan to investigate it.

- *Step 4: Create—Practice of developing and using models*

Students will obtain, evaluate, and communicate information by constructing explanations and designing solutions (SEP 8; MP 3). Students will develop and use models (SEP 2; MP 4). Students will build a prototype.

- *Step 5: Test and improve—Practice of obtaining, evaluating, and communicating information*

Students will analyze and interpret data to draw conclusions and apply understandings to new situations (SEP 4; MP 5). Does the prototype work, and does it solve the need? Students communicate the results and get feedback and analyze and talk about what works, what does not work, and what could be improved.

One of the easiest ways we have found to demonstrate the engineering process is merely by putting a twist on the common paper airplane challenge that most of us are familiar with. We start off by helping students make a paper airplane. In a predetermined test area within the classroom, students verify that their planes can fly. Now is when you provide the challenge. Students are given a starting point to launch their airplanes. They are also given a landing zone. Next they are given the path the plane must take before it can land in the predetermined area. Most of the time, we keep this simple and have the students modify their plane so it can make a gentle curve. Students will quickly see that their first attempts did not work. Using what they learned, they will now make modifications to their original design in order to get the appropriate outcome. Although this task may seem simple, it does provide the students the understanding about engineering design they need in order to be successful on these next lessons. Again, we have learned the hard way not make assumptions about what students know.

Teacher briefs precede each lesson to explain our vision for each lesson with regard to its instructional target and also to provide some helpful tips to be successful with the lesson. In general, our lessons are provided as examples of how to take your existing lessons and turn them into STEM lessons by inserting an engineering design challenge. Our lessons are aligned to the *Framework* (NRC 2012), but this method would work with any set of standards. Individual lessons typically only target a portion of a larger concept. Those specific targets will be described within the teacher briefs. We have included a direct URL link to aligned CK-12 FlexBooks if additional background knowledge about a specific topic for students (and teachers) is needed.

All of the lessons have been tested by classroom teachers or instructional coaches. Feedback has been very positive regarding student learning and especially participation, which is music to our ears. In most cases, these lessons were tested in our district's most challenging schools (those with more than 70% of students receiving free and reduced-price lunch). As expressed previously, we are believers that ALL students must experience STEM, and perhaps the most deserving of highly engaging science lessons are these students living in high-poverty areas.

We hope you have as much fun implementing the lessons as we had designing them.

References

Cheuk, T. 2013. Relationships and convergences among the mathematics, science, and ELA practices. Refined version of diagram created by the Understanding Language Initiative for ELP Standards. Palo Alto, CA: Stanford University.

Danielson, C. 2013. *The framework for teaching: Evaluation instrument.* Princeton, NJ: The Danielson Group.

Marzano, R. J. 2007. *The art and science of teaching.* Alexandria, VA: ASCD.

Mayes, R., and T. R. Koballa, Jr. 2012. Exploring the science framework: Making connections in math with the Common Core State Standards. *Science and Children* 50 (4): 15–22.

National Research Council (NRC). 2012. *A framework for K–12 science education: Practices, crosscutting concepts, and core ideas.* Washington, DC: National Academies Press.

National Governors Association Center for Best Practices and Council of Chief State School Officers (NGAC and CCSSO). 2010. *Common core state standards.* Washington, DC: NGAC and CCSSO.

Teach Engineering. 2015. Engineering design process. *www.teachengineering.org/engrdesignprocess.php.*

CHAPTER
7 | Earth and Space Sciences StEMT Lessons

I love the added [engineering design] challenges to the lessons. It has taken time for the students to be successful but wow, just cool to see.

—Kelly at Lake Shipp Elementary

To Infinity and Beyond

Earth and Space Sciences: Earth's Place in the Universe—Earth and the Solar System (ESS1.B.)

Teacher Brief
Few topics are as difficult for students to understand as the relationships among the Sun, the Earth, and the Moon. Add the changing of constellations throughout the year, and you have your hands full as a teacher. It is for this very reason that we chose to include the engineering design challenge with a lesson involving Moon phases and the seasonal differences in how we observe constellations. Nothing about this concept is easy. For the learner, it requires something that few other concepts require—multiple perspectives. Therefore, we feel that the most important thing to consider before undertaking this lesson is that it will require time.

Knowledge of Moon phases is a requirement in most state science standards. Common activities include having students make observations of the night sky, walk around a lamp placed in the center of the classroom, or complete a diagram in small groups like we have included in this lesson. All of these endeavors have merit, and all activities may need to be completed for students to obtain the type of deep understanding for which we strive. We have included within the Explore section of this lesson a basic group investigation as a starting point. The students, depending on their experiences and grade level, may need more types of experiences before taking on the engineering challenge. It is assumed that some of the other common activities for Moon phases, listed above, have been completed with your students. With that said, here is what we have noticed when students attempt this challenge.

Most of our experiences with this lesson come from two distinct groups of participants. We have used this challenge with our coaches and teachers during professional development settings to help them understand what we believe to be the purpose of an engineering design challenge. In terms of the students, the classrooms that have participated in this challenge were Title I schools with close to 100% free and reduced-lunch status. What is interesting in watching these two distinct groups is that they both traverse the same path toward finding a solution. They also take approximately the same amount of time going through the steps. This is perhaps one of the important lessons to be learned for this engineering challenge: It does take time, regardless of experience and age level, and may require a minimum of 60–90 minutes for students (or teachers) to work through the evidence.

During the first 30 minutes of this challenge, groups will read and organize the provided data. Once they've organized the data, some groups will start to build a physical model, either with the cut-out clues or other objects. If you see groups not creating models, it should be encouraged. It may be as simple as asking a few questions of the groups and having them show you the positions. This is also when the teacher needs to be vigilant and force students back to what they have already learned in previous lessons on this content. If students use science notebooks, encourage them to have them out as a reference.

Once students start building their model, the lesson starts moving quickly toward a solution. The constellations, whether or not they are included in the state standards, are important, as they serve as cardinal directions on which other motions can be based. The constellations help the students in their initial model building to produce the correct placement of the planets. We would expect a teacher to scaffold as necessary, but know that fifth graders can successfully complete this lesson. We have also witnessed fifth-grade gifted students using a mathematical model to solve the problem. In either case, with a physical or a mathematical model, it is the modeling that provides the solution and should be stressed in the lesson debrief.

Given the time constraints that teachers face in teaching all the standards within a school year, at some point the teacher may need to end the lesson before each group has the correct answer. We do not see this premature conclusion as problematic. The goal is to have students develop a better understanding of the Earth in Space concept, not to find the solution.

By the end of grade 5. The orbits of Earth around the Sun and of the Moon around Earth, together with the rotation of Earth about an axis between its north and south poles, cause observable patterns. These include day and night; daily and seasonal changes in the length and direction of shadows; phases of the Moon; and different positions of the Sun, Moon, and stars at different times of the day, month, and year (*A Framework for K–12 Science Education* [*Framework*]; NRC 2012, p.176).

Framework Question: What are the predictable patterns caused by Earth's movement in the solar system?

StEMT*ified* Question: How many Planet J days will it take to align the planets and be closest to Earth in order to send a communication?

Dimension 1: Practices (What should students be doing?)

- Developing and using models (models of the Earth-Moon-Sun system; analyzing relationships to draw conclusions)

- Analyzing and interpreting data (explore and explain investigation)

- Engaging in argument from evidence (claims, evidence, reasoning)

- Obtaining, evaluating, and communicating information (comparing and contrasting similarities and differences in Earth-Moon-Sun system patterns)

Dimension 2: Crosscutting Concepts (*Framework*; NRC 2012, p. 84)

- *Patterns.* Observed patterns of forms and events guide organization and classification, and they prompt questions about relationships and the factors that influence them. Examples of patterns could include day and night; daily and seasonal changes in the length and direction of shadows; phases of the Moon; and different positions of the Sun, Moon, and stars at different times of the day, month, and year.

- *Systems and system models.* Defining the system under study—specifying its boundaries and making explicit a model of that system—provides tools for understanding and testing ideas that are applicable throughout science and engineering.

Dimension 3: Disciplinary Core Idea

The Earth and the Moon, Sun, and planets have predictable patterns of movement. These patterns, which are explainable by gravitational forces and conservation laws, in turn explain many large-scale phenomena observed on Earth (*Framework;* NRC 2012, p. 175). The Moon and Sun's positions relative to Earth cause lunar and solar eclipses to occur. The Moon's monthly orbit around Earth; the relative positions of the Sun, the Moon, and the observer; and the fact that it shines by reflected sunlight explain the observed phases of the Moon (*Framework;* NRC 2012, p. 175).

Framework Question: What are the predictable patterns caused by Earth's movement in the solar system?

Instructional Duration: Three to four 45-minute class periods

Materials per Class of 30

- Moon phases model sheet (see Appendix C, p. 193)
- Lamp or flashlight (one per group)
- Styrofoam ball (Sun or Moon; one per group)
- Constellation images (see Appendix B, p. 192)
- Solar system template
- Safety glasses or goggles

Safety Notes

1. Personal protective equipment (eye protection) is to be worn during the setup, hands-on, and takedown segments of the activity.

2. Use caution if using lamp. Bulb gets hot and can burn skin.

3. Keep lamp and cord away from water sources.

For additional information, teachers may use the OER CK–12 Standards-Aligned Flex-Book textbook *Earth Science* (*www.ck12.org/teacher*).

Guiding Question: Why does the Moon seem to change shape and why can we only see certain constellations at certain times of the year?

Misconceptions (adapted from AAAS Project 2061 [n.d.])

- The Moon can be seen only at night.

- The Moon rises at the same time the Sun sets.

- All planets take the same amount of time to orbit the Sun.

- A day is the same length on all planets.

- Galaxies are inside our solar system.

ENGAGE

- How will you capture students' interest and reveal misconceptions?

- What kinds of questions should the students ask themselves after the engagement?

In small groups, have students answer the following questions (do not confirm answers; continue to probe for clarification from students):

- Can you see the Moon at night? If so, describe how the Moon looks.

- Can you see the Moon every night? Why or why not?

- Can you see the Moon during the daylight hours? Why or why not?

- What about stars, can you see the same stars every night? Why or why not?

EXPLORE

- Describe the hands-on activities that will provide experience of the phenomenon.

- List "big idea" conceptual questions you will use to encourage and focus students' exploration and to allow the students to test their ideas.

Prior Knowledge

Constellations are groups of stars that form patterns in the night sky. These stars appear as small points of light because they are very far away and outside our solar system.

Explore 1

Students should have an understanding of the day and night cycle caused by Earth's rotation around its axis. Students must have an understanding that the Earth travels around (orbits) the Sun every 365 days and that this orbit is nearly circular.

1. Print out the Moon phases model sheet—one per student (Appendix C).

2. Place the flashlight so that it shines like the Sun's rays on the Moon phases model sheet.

3. Find the Earth in the center of the model sheet and place the Ping-Pong ball on it.

4. Shine the flashlight on the Ping-Pong ball (Earth) and determine daylight and nighttime.

5. Students color in the part of the Earth that is NOT illuminated on the Moon phase worksheet.

6. Students place the Ping-Pong ball in the circle in between the Earth and the Sun and shade in the part that is NOT illuminated. Students label the Moon phase.

7. Students place the Ping-Pong ball in the circle behind Earth and farthest away from the Sun and shade in the part that is not illuminated (full Moon). (*Note*: Students may have the misconception that the Earth always blocks the Sun's rays and causes an eclipse if the Earth is between the Sun and the Moon.)

8. Students complete the rest of the diagram, shading the parts of the Moon that are NOT illuminated. Students are not required to know the names of the phases of the Moon.

Explore 2

Place pictures of constellations around the room that correspond with what can be seen in the night sky. Have students align their Moon phase diagrams to the same direction (Appendix B). Working in small groups, students work together to answer the following questions in their journals:

• During the night, what constellation will you see overhead around midnight?

• What constellation can be seen right before the Sun rises in the morning sky?

- What constellation can be seen right after the Sun sets?

- What constellation can you not see in the night sky? Why?

Have students turn their data collection sheets 90°. Tell students that the Earth has now traveled a quarter of the way around the Sun. For example, instead of summer it is now fall.

- Can you see different constellations? Why?

- Which constellations can you see and which ones can't you see?

Look-Fors: students should (practices and behaviors):

- Use models of Earth-Moon-Sun system (SEP 2/MP 4).

- Be actively engaged and work cooperatively in small groups to complete investigations.

- Test solutions to problems, and draw conclusions.

- Use effective communication skills (writing, speaking and listening; SEP 7, SEP 8; MP 3).

- Analyze and interpret data to draw conclusions and apply understandings to new situations (comparing and contrasting similarities and differences in Earth–Moon–Sun system patterns; SEP 4; MP 5).

- Apply scientific vocabulary after exploring a scientific concept (SEP 6; MP 7).

EXPLAIN

- Student explanations should precede your explanations or introduction of terms. What questions or techniques will you use to help students connect their exploration to the concept under examination?

- List higher-order thinking questions that you will use to solicit *student* explanations and to help students construct and justify their explanations.

At this point in the lesson, students should be doing the explaining, and all answers should target the guiding question: *Why does the Moon seem to change shape, and why can we only see certain constellations at certain times of the year?* Ask students the following probing questions for small group discussion and random-call report out by students:

- Does the Moon make its own light? Where does moonlight come from?

- How much of the Moon is always illuminated by the Sun?

- When the Moon was directly between the Earth and the Sun, why could people on Earth not see the Moon easily?

- What is this kind of Moon called? Why?

- When did Earth "see" a full Moon?

- How long does it take for the Moon to move through a complete lunar cycle?

- Why does the Moon seem to have different shapes?

Students answer the following question in their notebooks: The Earth and the Moon, Sun, and planets have predictable patterns of movement. Explain why the revolution of the Earth around the Sun causes different constellations to be seen at different times of the year.

ELABORATE

- Describe how students will develop a more sophisticated understanding of the concept.

- What vocabulary will be introduced, and how will it connect to students' observations?

- How is this knowledge applied in our daily lives?

Learn more about the Moon in *The Moon Book* by Gail Gibbons. Students can research to learn the difference between a solar eclipse and a lunar eclipse. Have them sketch the Earth, the Sun, and the Moon and explain how they are aligned during an eclipse.

StEMT*ify*

Science: What real-world issues and problems exist in the area of space exploration?

Engineering: Is the STEM lesson guided by the engineering design process?

- Students should be immersed in hands-on inquiry and open-ended exploration.
- Students should be involved in productive teamwork.

Mathematics: Does the STEM lesson rigorously apply mathematics and science content that the students are learning?

Technology: Does the STEM lesson allow for multiple right answers through prototype development, testing, and improvement of design? Encourage students to monitor and evaluate their progress, change course if necessary, and ask, "Does this make sense?"

Guiding Question: Why does the Moon seem to change shape, and why can we only see certain constellations at certain times of the year?

StEMT*ified* Question: How many Planet J days will it take to align the planets and be closest to Earth in order to send a communication? Explain your answer.

Preparation: Students will be provided with the prereading text shown in Table 7.1 (p. 46). This text provides the background for the challenge. It also holds many of the pieces of evidence students will need to develop a resolution to the problem. It is important that students read carefully and mark the text where critical information is contained.

Step 1: What is the problem to solve? What is the goal?

Students will be provided with an envelope containing the evidence cards (one set per group). The evidence cards are marked as they pertain to Planet M or J. Students will combine the evidence from the cards (evidence gathered from the ship's failing computer system) and develop a plan. Optional: Provide students with the graphic organizer (solar system model template).

Table 7.1. Preactivity reading for Elaborate activity for Lesson ESS1.B

You look out the spacecraft's tiny window. There is not much to see really—a whole lot of darkness. Being an astronaut is not all that glamorous. It has been months since you have even spoken to another human being. It will be nice to get to Planet J, the science station in a nearby solar system. The station gets resupplied on a regular basis, but this is your first trip there. You look at the onboard computer and it shows 18 more hours until you land on Planet J.

As solar systems go, this system is very similar to your own. It has one star and several planets that orbit the star. Several of the planets have small moons. Coming into the solar system from a magnetic north orientation, all the planets orbit the star in a counterclockwise manner. All the moons rotate in a counterclockwise motion as well.

Bam! "What was that?" you wonder. You look at the gauges and you notice that the capsule's air pressure is dropping. It must have been a micrometeorite hitting the ship. As fast as you are going, something the size of a grain of sand can go straight through the ship. It's a good thing this next-generation ship has self-sealing skin. You see the air pressure gauge stabilize and start to go back up. It's time to check to see if there was any damage. You go through your prepared checklist. About halfway through the list, you notice a problem with the spaceship's computer system. You need to get this bird down now while you have a chance to land safely. You notice you are headed toward Planet M, a planet in the same solar system as Planet J. There is no way to get to Planet J with these damages—all you can do is attempt to land safely on Planet M.

You land safely on Planet M, another planet in the solar system. You have enough food, water, and air for 270 days based on your home planet's 24-hour day. Because of the damage to your computer system, you have lost most of the information about this planet and its relationship to Planet J. You have one shot at being rescued off Planet M. Your backup radio only has enough signal strength to reach Planet J when the planets are very close. You also only have enough battery power to send one message. You have one shot to make this count! You need to figure out when the planets will be in close proximity to one another.

Fortunately, you were able to retrieve a little bit of information from the computers before they stopped working. However, there is one critical question that must be answered: *How many Planet J days will it take to align the planets in order to send a communication?*

Step 2: Research and design—Practice of planning and carrying out investigations

Students collaborate to brainstorm ideas and develop as many solutions as possible. Although not absolutely necessary, it might be a good idea for students to present their findings and their analytical approach to the class before they move on to the next step of the lesson.

Step 3: Plan—Practice of constructing explanations and designing solutions

Students will compare the best ideas, select one solution and make a plan to investigate it. During the lesson, it is important to probe students' thinking. Most will still struggle with the concepts of day and night, moon phases, and locations of galaxies. Encourage them to construct a model based on their evidence. The key to this problem-solving activity is for students to design a model. Once constructed, students will need to manipulate the model to find the solution to the problem. Given the complexity of this problem, it will take students many attempts at manipulating their model before they can construct a viable solution. The important thing to consider for this activity is that it is more important for students to be able to manipulate the model than to provide an exact answer.

Step 4: Create—Practice of developing and using models

Students will build a model.

Step 5: Test and improve—Practice of obtaining, evaluating, and communicating information

Does the model work, and does it solve the need? Students communicate the results and get feedback. Students then analyze and talk about what works, what does not, and what could be improved.

EVALUATE

- How will students demonstrate that they have achieved the lesson objective?

- Evaluation should be embedded throughout the lesson as well as at the end of the lesson.

Summative Assessment: Write about it!

Claim: Write a sentence stating how many Planet J days it will take to align the planets and send a communication.

Evidence: Provide scientific data from the activity to support your claim. The evidence should include how the number of days was determined.

Reasoning: Explain why your model supports your claim.

- Differentiation could include varied product formats, such as oral presentation, poster presentation for a gallery walk, or storyboard.
- Refer to the Claims, Evidence, Reasoning (CER) Rubric in Appendix A.
- Ask students to use evidence from the StEMTify activity to elaborate on the answer for the following question: Why does the Moon seem to change shape and why can we only see certain constellations at certain times of the year?

Oh, How the Garden Grows

Earth and Space Sciences: Earth's Systems—Earth Materials and Systems (ESS2.A)

Teacher Brief

For this disciplinary core idea, we chose a rather simple lesson focused on soils and the role soils play in plant growth. Depending on the scope and sequence, this lesson may be preceded by other lessons that involve soil development. The lesson we chose starts with students being asked questions during the Engage portion about what plants need. This is when the teacher can experiment or just ask the basic questions provided. There is no right or wrong question or answer; there is only what works to initiate student learning.

The Explore portion has students making observations to conclude that soils are a mixture of materials and that there are many different types of soils with different characteristics. This Explore and Explain sections could easily be modified to allow for a more in-depth study of soils and soil types as needed for specific state standards.

The engineering design challenge we have included uses one attribute of soils (water retention) as the basis for the problem. We encourage the teacher to modify the attribute, such as resistance to erosion, to meet specific standards. Students are provided with materials they will use to form a mixture that they feel will have a similar characteristic to a soil that retains water in sufficient amounts for a plant to grow. The materials we used with the students were chosen because they were readily available to teachers. There is nothing magical about our list of supplies, so please feel free to experiment. In one case, a teacher extended the lesson by tossing some seeds into the students' mixtures. The class observed the growth over the next few weeks to integrate practice of science concepts.

By the end of grade 5. Earth's major systems are the geosphere (solid and molten rock, soil, and sediments), the hydrosphere (water and ice), the atmosphere (air), and the biosphere (living things, including humans). These systems interact in multiple ways to affect Earth's surface materials and processes. The ocean supports a variety of ecosystems and organisms, shapes landforms, and influences climate. Winds and clouds in the atmosphere interact with the landforms to determine patterns of weather. Rainfall helps shape the land and affects the types of living things found in a region. Water, ice, wind, living organisms, and gravity break rocks, soils, and sediments into smaller particles and move them around. Human activities affect Earth's systems and their interactions at its surface (*Framework;* NRC 2012, p. 181).

Framework **Question:** How do Earth's major systems interact?

StEMT*ified* Question: For agricultural purposes, why is it important to enhance sandy soils?

Dimension 1: Practices (What should students be doing?)

- Developing and using models (interactions in the geosphere, biosphere, hydrosphere, and atmosphere)

- Analyzing and interpreting data

- Engaging in argument from evidence to explain a phenomenon (effects of weathering or the rate of erosion by water, ice, wind, or vegetation)

Dimension 2: Crosscutting Concepts

- *Energy and matter. Flows, cycles, and conservation.* Tracking fluxes of energy and matter into, out of, and within systems helps students understand the systems' possibilities and limitations.

- *Cause and effect.* Cause-and-effect relationships are routinely identified, tested, and used to explain change.

- *Systems and system models.* A system can be described in terms of its components and their interactions.

Dimension 3: Disciplinary Core Idea

Earth is a complex system of interacting subsystems: the geosphere, hydrosphere, atmosphere, and biosphere. The geosphere includes a hot and mostly metallic inner core; a mantle of hot, soft, solid rock; and a crust of rock, soil, and sediments. The atmosphere is the envelope of gas surrounding the planet. The hydrosphere is the ice, water vapor, and liquid water in the atmosphere, ocean, lakes, streams, soils, and groundwater. The presence of living organisms of any type defines the biosphere; life can be found in many parts of the geosphere, hydrosphere, and atmosphere. Humans are of course part of the biosphere, and human activities have important effects on all of Earth's systems. All Earth processes are the result of energy flowing and matter cycling within and among Earth's systems. This energy originates from the Sun and from Earth's interior. Transfers of energy and the movements of matter can cause chemical and physical changes among Earth's materials and living organisms (*Framework;* NRC 2012, p. 179).

Instructional Duration: Three to four 45-minute class periods

Materials

Materials per class: Bags of three or four different soil types gathered from different locations or purchases (e.g., sand or a sandy soil; a more fertile soil with organic matter such as a potting soil; and a hard, clay-like soil). *Note:* Clay soils can be found at sand and gravel manufacturers. Another source you may want to check with before you purchase this item is your school district's athletic director, as it is used on baseball fields. Other ideas: corn starch (clay soil substitute), dry mulch bark, or pea gravel.

[Teacher safety note: Make sure you know the source of the soil you use. Some soils contain pesticides, herbicides, broken glass, and so on. It's safer to buy commercially available soil types that have been sanitized.]

Materials per group: 250 ml beaker, 50 ml graduated cylinder, piece of screen, forceps or toothpicks, several Styrofoam cups, tray with shallow sides (a cafeteria tray can work for this), paper plates, coffee filter paper or ground cloth.

Per Person: Indirectly vented chemical splash goggles, nonlatex gloves, nonlatex apron.

Safety Notes

1. Personal protective equipment (splash goggles, gloves, and aprons) should be worn during the setup, hands-on, and takedown segments of the activity.

2. Use caution when working with sharps (toothpicks, forceps, etc.) that can cut or puncture skin.

3. Wash hands with soap and water upon completing this activity.

Misconceptions (adapted from AAAS Project 2061)

- Catastrophic changes on the Earth's surface, like volcanic eruptions and earthquakes, only affect the lithosphere.

- The atmosphere, hydrosphere, lithosphere, and biosphere do not cause changes in one another; these systems operate independently on Earth.

ENGAGE

- How will you capture students' interest and reveal misconceptions?
- What kinds of questions should the students ask themselves after the engagement?

In small groups, have students discuss the following probing questions:

- Have you ever grown a plant from a seed? What do plants need to grow?
- What might happen to plant growth in an area after a volcanic eruption?
- What role does soil play in plant growth? How might a volcanic eruption affect soils?
- Why do scientists study soils (pedology)?

The water we drink, the air we breathe, and the soil we use for agricultural purposes are three of our most important natural resources.

EXPLORE

- Describe the hands-on activities that will provide experience of the phenomenon.
- List "big idea" conceptual questions you will use to encourage and focus students' exploration and to allow the students to test their ideas.

Prior Knowledge
Students should have previously learned how to read and interpret a table. They also should have learned in earlier grades about the needs and parts of plants.

Background Information
Soil is an integral part of the ecosystem. There are many soil types on Earth. Soil is made up of sediment from the weathering and erosion of rock and organic matter. The soil type of an area is a direct result of the available rocks and organic matter of the area. Some areas, such as Florida, have very sandy soils. Other areas, such as the Midwest, have soils consisting of clay and silt. The ability of plants to grow is related to the type of soil.

For additional information on soils, teachers may use the OER CK–12 Standards-Aligned FlexBook textbook, *Earth Science (www.ck12.org/teacher)*.

Lesson Procedures

- Students will be supplied with three to four different soil types gathered from different locations. Preferably, one should be a sample of very sandy soil, the second a sample of fertile soil, and the third a sample of hard, clay-like soil.

- Students (in small groups) will pour their samples onto paper plates. Students will use a toothpick and a small hand lens to examine the soil contents.

Phase 1

In small groups, have students answer the following questions as they observe their soils. They should record their observations in their notebooks using the chart in Table 7.2.

1. Using senses: What do you see in the soil sample? How does it feel? How does it smell?

2. How do you think your soil sample formed?

   ○ Identify organic material (leaves, twigs) that have decomposed

   ○ Identify pieces of larger rocks. How do large rocks become smaller rocks? (*Teacher may need to review process of weathering*)

3. Which kind of soil would be best for plant growth? Why?

4. Which soil do you think will hold the most water? Why? Is this important? Why?

Table 7.2. Soils data chart for Lesson ESS2.A

Type of Soil	Color (e.g., dark brown, light brown, reddish brown)	Texture (e.g., gritty, sandy, or smooth)	Water Retention (e.g., good, poor, excellent)	Types of Material in Soil (e.g., rocks, twigs, shells)	Ability to Grow Plants (yes or no)
Soil Sample 1					
Soil Sample 2					
Soil Sample 3					
Soil Sample 4					

5. What could be added to the soil to allow it to hold more water?

6. Can a plant have too much water? Why?

Phase 2

Students take their three to four soil samples. They will put one of the samples in a strainer and hold it over a pan. With a dropper, carefully drip water onto the soil sample to see how well it absorbs the water. (Students will need to drop the water on the soil slowly to prevent it from running off the surface. Students will record their findings in their data collection table.)

Phase 3

Choose three very different types of soil from the samples brought to class by the students. Have students plant flower or vegetable seeds in each type of soil, water them and place them in the sunlight. Have the students continue to water the plants over the course of a couple of weeks, and monitor the development of the seeds into plants. Ask the students to determine if one kind of soil was better for plant development than the others, and ask why they believe this was the case. Students will be provided with the first data set displaying the attributes of different soils.

EXPLAIN

- Student explanations should precede your explanations or introduction of terms. What questions or techniques will you use to help students connect their exploration to the concept under examination?

- List higher-order thinking questions that you will use to solicit *student* explanations and to help students construct and justify their explanations.

After all of the teams have completed their data table and observed plant growth, each student should prepare a report explaining what her or she learned about soils. The report should include an argument for which type of soil is better and why.

ELABORATE

- Describe how students will develop a more sophisticated understanding of the concept.

- What vocabulary will be introduced and how will it connect to students' observations?

- How is this knowledge applied in our daily lives?

An elaboration for this lesson might be to do a close read on a text about the loss of soil fertility. After reading, students might write a summary of what they learned in their science notebooks.

StEMT*ify*

Science: What real-world issues and problems exist in the area of pedology (study of soils) and the effect on way of life?

Engineering: Is the STEM lesson guided by the engineering design process?

- Students should be immersed in hands-on inquiry and open-ended exploration.

- Students should be involved in productive teamwork.

Mathematics: Does the STEM lesson rigorously apply mathematics and science content that the students are learning?

Technology: Does the STEM lesson allow for multiple right answers through prototype development, testing, and improvement of design? Encourage students to monitor and evaluate their progress, change course if necessary, and ask, "Does this make sense?"

StEMT*ified* Question: For agricultural purposes, why is it important to enhance sandy soils?

Engineering Design Challenge

Background knowledge: Florida soils are very sandy. As a result, they do not hold moisture. Even though Florida receives an abundance of rainfall, crops struggle to stay hydrated. This need for water is met by an abundance of irrigation systems. It would be far better for the environment if not as much water was needed to irrigate crops and lawns. The question that must be asked, is there a cheap, simple solution that can help Florida soils absorb water? Keep in mind that soils that stay saturated all the time can place the plants that grow in these soils at risk of rotting.

Step 1: Ask—Practice of asking questions (science) and defining problems (engineering)

What is the problem to solve? What needs to be designed, and who is it for? What are the project requirements and limitations? What is the goal?

Students will be given a set of materials to use that may or may not retain water. They will use these materials to design a soil that will hold water the longest but will not stay saturated. Saturation will mean that 90% of the water is retained in the soil and does not pass through into the capture beaker. The goal is to get the soil sample to hold as much water as possible, not to exceed the 10% threshold.

Step 2: Research and design—Practice of planning and carrying out investigations

Students collaborate to brainstorm ideas and develop as many solutions as possible.

Step 3: Plan—Practice of constructing explanations and designing solutions

Students will compare the best ideas, select one solution and make a plan to investigate it. Students will be provided with three to four small cups of dry sand. These will be used to prepare their test soils. Students will also be provided with materials that can be mixed into the soil. Students should do an initial test on these materials to see how well a small sample will absorb water.

Step 4: Create—Practice of developing and using models

Students will build a prototype.

1. Students will place a small piece of coffee filter in the bottom of a paper or Styrofoam cup and poke seven or eight holes into the bottom of the cup to allow drainage. Students will then place their carefully designed mixture into the cup.

2. Students will hold their test soil over a large baking dish to prevent a mess from being made. A small beaker will be placed under the cup holding the test soil. One student will hold the test soil cup above the beaker.

3. One of the group members will slowly pour 50 ml of water into the sample.

4. Once all the water has come out of the cup, the volume of liquid that has passed through the soil sample will be recorded from the capture beaker.

5. Students will calculate the differences between the water that was poured into the cup and what runs out and is captured by the beaker. The difference between what is poured in and what is captured will be the amount absorbed by the soil.

Step 5: Test and improve—Practice of obtaining, evaluating, and communicating information

Does the prototype work, and does it solve the need? Students communicate the results and get feedback and then analyze and talk about what works, what does not, and what could be improved.

Using the knowledge gained in the first trial, the students will use the engineering design cycle to improve their soil. Repeat for Trial 3. (See Table 7.3.)

Table 7.3. Engineering design process data chart for Lesson ESS2.A

Trial	Contents of Mixture (g)	Amount of Water (Start)	Amount of Water (After)	Amount of Water Absorbed by Soil
Trial 1				
Trial 2				
Trial 3				

For an extension activity, students can experiment with the soils they created and determine if plants can grow in the engineered soils.

EVALUATE

- How will students demonstrate that they have achieved the lesson objective?
- Evaluation should be embedded throughout the lesson as well as at the end of the lesson.

Summative Assessment: Write about it!

Students will write a report to Florida government officials describing their design and how it might affect soil quality and water management in Florida using the *Claims, Evidence, Reasoning* format. *Why is it important to enhance sandy soils? Include how rainfall helps to shape the land and affects the types of living things found in a sandy climate region.*

Refer to Appendix A for the CER Rubric.

Separation Anxiety

Earth and Space Sciences: Earth and Human Activity—Natural Hazards (ESS3.B.)

Teacher Brief

What is not to like about disaster preparedness lessons? They provide students with real-world relevance and high interest. We start this lesson with a basic disaster preparedness lesson, which can certainly be adapted to other areas of the country. Students are asked to think about items their family may need for several days. It is important to note; this type of thinking can be difficult for students, as there is no one correct answer. For those teachers with a strong belief that these types of lessons would be a great opportunity to bring in some mathematics, the lesson is easily modifiable to include a budget for supplies that students would need to purchase.

The engineering design challenge brings in the hands-on component that is too often missing in disaster preparedness lessons. We chose to create this challenge using very simple materials, but it can be modified using other materials available to the teacher. Other substitutes for this challenge may be a shake-table test, erosion prevention, or even water waves if your location has such hazards. This lesson, designed originally to meet the requirements for a fifth-grade standard, can also be increased in difficulty and complexity for higher grades using different materials and design constraints.

By the end of grade 5. A variety of hazards result from natural processes (e.g., earthquakes, tsunamis, volcanic eruptions, severe weather, floods, coastal erosion). Humans cannot eliminate natural hazards, but we can take steps to reduce their effects.

Framework Question: How do natural hazards affect individuals and societies?

StEMT*ified* Question: How can roof structure be improved to reduce the amount of storm damage during high-wind events?

Dimension 1: Practices (What should students be doing?)

- Developing and using models (design roof structures)

- Analyzing and interpreting data (compare and contrast the stability of different roof structures)

- Engaging in argument from evidence (make a claim about the merit of a design solution that reduces the impacts of a weather-related hazard)

- Obtaining, evaluating, and communicating information (generate and compare multiple solutions to reduce the impacts of natural Earth processes on humans)

Dimension 2: Crosscutting Concepts

- *Cause and effect: Mechanism and explanation.* Events have causes and cause and effect relationships are routinely identified, tested, and used to explain change. Such relationships can then be tested across given contexts and used to predict and explain events in new contexts.

- *Systems and system models.* Defining the system under study—specifying its boundaries and making explicit a model of that system—provides tools for understanding and testing ideas that are applicable throughout science and engineering.

- *Stability and change.* For natural and built systems alike, conditions of stability and determinants of rates of change or evolution of a system are critical elements of study.

Dimension 3: Disciplinary Core Idea

Natural processes can cause sudden or gradual changes to Earth's systems, some of which may adversely affect humans. Through observations and knowledge of historical events, people know where certain types of these hazards—such as earthquakes, tsunamis, volcanic eruptions, severe weather, floods, and coastal erosion—are likely to occur. Understanding these kinds of hazards helps us prepare for and respond to them. (*Framework;* NRC 2012).

Instructional Duration: Three to four 45-minute class periods

Materials

Teacher materials: chart paper, internet access, projector, document camera, digital camera

Materials per group: modeling clay, paper (several pieces), masking tape (one meter per trial), floor fan or hair dryer (cool temperature setting), craft sticks (6), toothpicks (15), two-liter plastic bottle, scissors

Materials per student: Student notebook, pencil, indirectly vented chemical splash goggles, nonlatex gloves

Safety Notes

1. Personal protective equipment (splash goggles, gloves, and aprons) is to be worn during the setup, hands-on, and takedown segments of the activity.

2. Use caution when working with sharps (sticks, toothpicks, scissors, etc.) that can cut or puncture skin.

3. Use caution when working with electrical appliances (fan, hair dryer, etc.). Keep away from water sources to prevent shock hazard.

4. Wash hands with soap and water upon completing this activity.

For a basic overview of natural hazards, teachers may use the OER CK–12 Standards-Aligned FlexBook textbook *Earth Science* (*www.ck12.org/teacher*). Also, learn about other types of disasters at *www.ready.gov/natural-disasters*.

Guiding Question: How can I help my family prepare for natural disasters?

Misconceptions

- Natural disasters happen very rarely, and these events are just the bad luck of the people that are affected.

- Floods are rare, atypical, almost unnatural events rather than normal river behavior.

- All natural disasters have only local effects.

- Flooding only occurs after a heavy rainfall.

- Global warming is caused by the hole in the ozone because it lets in more radiation.

- Ash from a faraway volcano cannot affect, Florida, Michigan, Virginia, and so on.

Ask students the following probing questions:

- What are some possible natural disasters that can affect people?
- What are some possible items a family would need if a disaster struck in their area and they had to evacuate or if they were stuck in their homes without power?
- What are five items you could not live without?

Have students share their list and discuss how their families may have different needs.

Prior Knowledge

In previous grades, students should have learned about the importance of being prepared for severe weather.

1. In groups, have students create a list of the items they think should be in their family's disaster kit. In other words, what would they need to survive for three to five days? Encourage students to debate with their peers and provide evidence for their choices. Each team must come to a consensus on a list.

2. Student groups should then rank their items from most important to least important. Again, each group must come to a consensus.

3. Have student groups list their items on chart paper and display the papers around the room.

4. Allow students time to review the choices of their peers.

5. After the gallery walk, provide time for each student to develop his or her own list.

6. Students should take the list home to discuss it with their parents.

7. Students should also discuss the following issues with their parents:

- Where to meet away from home in case of a fire

- Where to meet outside the neighborhood if you must evacuate

- Where to call to "check in" if you become separated from your family during a disaster

Students should memorize the phone number of a family member who lives in another state to report where they are so their family can find them.

At this time, you may want to give students an assignment sheet that can be returned at a later date with a parent's signature. On the assignment sheet, you may want to include the student's prioritized list to include in their emergency kit.

EXPLAIN

- Student explanations should precede your explanations or introduction of terms. What questions or techniques will you use to help students connect their exploration to the concept under examination?

- List higher-order thinking questions that you will use to solicit *student* explanations and to help students construct and justify their explanations.

Students should bring their project (or present their potential project) on the planned due date to share with the class.

1. Classmates should compare and contrast emergency plans. Is there a one-size-fits-all plan?

2. Why would one family's plan look different from another?

3. Can the list change based on the climate or season in which the disaster happens? If so, in what ways?

4. Can the list be different depending on the type of disaster? (For example, a power outage from a storm compared to a flood.) If so, in what ways?

Host a roundtable discussion on why students and families should have disaster plans.

ELABORATE

- Describe how students will develop a more sophisticated understanding of the concept.

- What vocabulary will be introduced, and how will it connect to students' observations?

- How is this knowledge applied in our daily lives?

Give each group of students a natural disaster scenario (fire, tornado, hurricane, blizzard, etc.). Have students write a story about a family that is prepared. The story must include the items they have included in their plan or any new items they decide they may need.

StEMT*ify*

Science: What real-world issues and problems exist in the area of hazardous weather or natural disasters?

Engineering: Is the STEM lesson guided by the engineering design process?

- Students should be immersed in hands-on inquiry and open-ended exploration.
- Students should be involved in productive teamwork.

Mathematics: Does the STEM lesson rigorously apply mathematics and science content that the students are learning?

Technology: Does the STEM lesson allow for multiple right answers through prototype development, testing, and improvement of design? Encourage students to monitor and evaluate their progress, change course if necessary, and ask, "Does this make sense?"

StEMT*ified* Question: How can roof structures be improved to reduce the amount of storm damage during high-wind events?

Student Background

Hurricane Andrew hit the Miami area in Florida on August 24, 1992. It made landfall as a Category 5 hurricane. Winds from the storm reached speeds in excess of 175 mph (280 km/h). The damage caused by Hurricane Andrew prompted new building codes that would better protect buildings against such powerful future storms. The question that was asked by engineers was, how do you design a house that can withstand such severe winds? The majority of wind damage is caused when air is able to enter a structure. On many occasions, homes in the Miami area had their roofs lifted off. Some types of roof structures survived more often than others. So, how can a roof structure be built to minimize the effects of strong sustained winds?

Teacher Note

Students may not be familiar with different types of roofs. A quick slide presentation of roof types might be in order before proceeding with this lesson.

Step 1: Ask—Practice of asking questions (science) and defining problems (engineering)

What is the problem to solve? What needs to be designed, and who is it for? What is the goal?

Step 2: Research and Design—Practice of planning and carrying out investigations

Students collaborate to brainstorm ideas and develop as many solutions as possible to design a roof structure that can withstand a sustained wind without loss of the structure (breaking off). What are the project requirements and limitations?

Step 3: Plan—Practice of constructing explanations and designing solutions

Students will compare the best ideas, select one solution, and make a plan to investigate it.

Step 4: Create—Practice of developing and using models

Students will build a clay structure that is six inches wide, six inches long, and three inches tall. Place craft sticks along the top rim of the clay structure. This will prevent the roof structure from sticking to the walls in the wind test. This will be the test structure (Figure 7.1).

Figure 7.1 Test structure

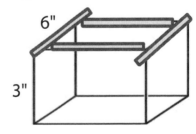

Students will use sheets of paper, approximately one meter of masking tape (per trial), six craft sticks, one two-liter plastic bottle, and 15 toothpicks to design a roof structure capable of withstanding high winds. During this stage, students should be given a time limit for construction. Students will carefully place their roof structure on the test building. The teacher will turn on a floor fan (to low)

that should be aimed directly at the model to replicate the constant hurricane winds. If the roof does not blow off, the teacher will turn up the fan to the next level. The teacher will repeat this action until the fan is blowing at full force.

Safety Notes

1. Personal protective equipment (splash goggles) should be worn during the setup, hands-on, and takedown segments of the activity.

2. Use caution when working with sharps (sticks, toothpicks, scissors, etc.) that can cut or puncture skin.

3. Use caution when working with electrical appliances (fan, etc.). Keep appliances away from water sources to prevent shock hazard.

4. Wash hands with soap and water upon completing this activity.

Step 5: Test and improve—Practice of obtaining, evaluating, and communicating information

Does the prototype work, and does it solve the need? Students communicate the results and get feedback. Then they talk about what works, what does not, and what could be improved.

Students will notice how and where their roof structure failed. They will take their design back to their design studio. They will make the necessary changes based on what they learned from their first test. Once finished, they will repeat the wind test.

Look-Fors

Students should engage in the following practices and behaviors:

- *Practice of asking questions (science) and defining problems (engineering).* Students will ask questions, define problems, and predict solutions/results (SEP 1; MP 1).

- *Practice of planning and carrying out investigations.* Students will be actively engaged and work cooperatively in small groups to complete investigations, test solutions to problems, and draw conclusions. Use rational and logical thought processes, and effective communication skills (writing, speaking and listening) (SEP 7, SEP 8; MP 3).

- *Practice of constructing explanations and designing solutions.* Students will design, plan, and carry out investigations to collect and organize data (SEP 3; MP 1).

- *Practice of developing and using models.* Students will obtain, evaluate, and communicate information by constructing explanations and designing solutions

(SEP 8; MP 3). Students will develop and use models of different types of roof structures) (SEP 2; MP 4).

- *Practice of obtaining, evaluating, and communicating information.* Students will analyze and interpret data to draw conclusions and apply understandings to new situations (SEP 4; MP 5). Apply scientific vocabulary after exploring a scientific concept (SEP 6; MP 7).

EVALUATE

- How will students demonstrate that they have achieved the lesson objective?
- Evaluation should be embedded throughout the lesson as well as at the end of the lesson.

Summative Assessment: Write about it!

Students write to explain why planning for disaster is so important. *How do natural disasters affect individuals?* Students should make a claim as to why it is important to plan for natural disasters, from developing building codes to prevent property loss to having a plan for your family in case the unthinkable happens. Students should support their claim through the use of evidence from both the Explore and Stemt*ify* sections of this lesson. Students should also explain how a roof blowing off their house will change how or where their family might live and how they survive for the following days (as compared to just losing power) using the Claims, Evidence, Reasoning protocol.

Students will create a brief presentation about their roof design (one piece of chart paper or no more than two PowerPoint slides). Groups will present their findings about the roof structures they tested using a claims, evidence reasoning format. They will make a claim as to why their proposed design was successful, provide evidence for this claim, and explain how this evidence supports their claim. The presentation should be professional as if it were being given to a group of home builders or to city building code officials.

See Appendix A for a sample CER rubric.

The Human Factor

Earth and Space Sciences: Earth and Human Activity— Human Effects on Earth Systems (ESS3.C)

Teacher Brief

Understanding environmental stewardship is a critical component of scientific literacy that we hope all students will understand and embrace. Within the backdrop of a very typical mixtures lesson, students will develop an understanding of what mixtures are and that they can be separated. This latter portion provides the opportunity for students to interact with the possibilities that the effects of human activity are reversible with informed and responsible management. Students will see throughout the lesson that some materials in mixtures are more easily separable than others. In a real-world setting, the difficulty of separating mixtures may greatly affect the cost of cleanup, and knowing this fact can encourage conservation and prevention.

The engineering design challenge we have chosen calls for students to develop a technology or technologies to separate the materials in a mixture. Devising a means of separation for each type of material will expose the students to the challenges of separating different types of materials. In some cases, it may not be possible for students to separate individual types of materials. It is then the teacher's option to modify the lesson and use it as a teaching opportunity to discuss why it is difficult to separate certain materials and what this means for real-world scenarios. Again, this challenge is about opportunities for learning the concept of human impact, not about developing the perfect design.

By the end of grade 5. Human activities in agriculture, industry, and everyday life have had major effects on the land, vegetation, streams, ocean, air, and even outer space. But individuals and communities are taking steps to help protect Earth's resources and environments. For example, we are treating sewage, reducing the amounts of materials we use, and regulating sources of pollution such as emissions from factories and power plants or the runoff from agricultural activities.

Framework Question: How do humans change the planet?

StEMT*ified* Question: How can we use knowledge of how mixtures can be separated as well as engineering design practices to design a solution to provide cleaner water to people around the globe?

Dimension 1: Practices (What should students be doing?)

- Developing and using models
- Analyzing and interpreting data
- Engaging in argument from evidence
- Obtaining, evaluating, and communicating information

Dimension 2: Crosscutting Concepts

- *Cause and effect: Mechanism and explanation.* Events have causes. Such mechanisms can then be tested across given contexts and used to predict and explain events in new contexts.
- *Systems and system models:* A system can be described in terms of its components and their interactions.

Dimension 3: Disciplinary Core Idea

Human activities in agriculture, industry, and everyday life affect the land, rivers, ocean, and air. Humans affect the quality, availability, and distribution of Earth's water through the modification of streams, lakes, and groundwater. Large areas of land, including such delicate ecosystems as wetlands, forests, and grasslands, are being transformed by human agriculture, mining, and the expansion of settlements and roads. Human activities now cause land erosion and soil movement annually that exceed all natural processes. Air and water pollution caused by human activities affect the condition of the atmosphere and of rivers and lakes, with damaging effects on other species and on human health. Some negative effects of human activities are reversible with informed and responsible management (*Framework;* NRC 2012, p. 194).

Instructional Duration: Four to five 45-minute class periods

Materials per Class of 30

Teacher materials: Assortment of mixtures, such as a box of crayons, deck of cards, trail mix, shells, rocks, etc.

For each group: Plastic cup with mixture of the following: six marbles, unpopped popcorn kernels, iron nuggets, fine-grained sand and coarse-grained sand, various sized screens for sifting, one tray, one magnet, five clear plastic cups (other objects can be substituted as needed)

Per student: Science notebook; pencil; safety goggles

Safety Notes

1. Personal protective equipment (splash goggles) should be worn during the setup, hands-on, and takedown segments of the activity.

2. Students are not to eat any food items used in this activity to avoid the risk of cross-contamination.

3. Wash hands with soap and water upon completing this activity.

For a basic overview of substances and mixtures, teachers may use the OER CK–12 Standards-Aligned FlexBook textbook *Physical Science* (*www.ck12.org/teacher*).

Guiding Questions: What is a mixture? How can a mixture be separated?

Misconceptions

- Earth and its systems are too big to be affected by human actions.
- If the Earth is too polluted for humans to live on, then we can move to another planet.
- The global Earth environment is stable relative to human timescales.
- The ozone hole is directly related to global warming.
- Human activities cannot affect geological processes like river flow, flood cycles, and so on.
- Technological fixes will save us from ruining our planetary environment.
- Coal burning is not an environmental problem because humans have invented ways to prevent pollutants from coal burning from entering the atmosphere.

ENGAGE

- How will you capture students' interest and reveal misconceptions?
- What kinds of questions should the students ask themselves after the engagement?

Display an assortment of mixtures, such as shells, crayons, or rocks. Ask:

- How are these materials alike?
- Can these materials be sorted and separated? How?
- Which materials would be easy to separate? Hard to separate?

Tell students that all of these assortments are mixtures. Have students write the key questions and any preliminary thoughts in their science notebooks. Provide students a few moments to talk this question over with a partner during this time. Allow student pairs to report out their thoughts (use this as a basis for the students' preliminary knowledge and misconceptions to help you guide your instruction).

EXPLORE

- Describe the hands-on activities that will provide experience of the phenomenon.
- List "big idea" conceptual questions you will use to encourage and focus students' exploration and to allow the students to test their ideas.

Prior Knowledge

Give each group their materials and tell them what is in the mixture. Ask students to find a way to separate the three materials and to write a sentence about how they did it in their science journals. After separating the mixture, have students place each of the different items into a separate cup after separating. Allow student groups to share their methods of separating the mixtures.

EXPLAIN

- Student explanations should precede your explanations or introduction of terms. What questions or techniques will you use to help students connect their exploration to the concept under examination?
- List higher-order thinking questions that you will use to solicit *student* explanations and to help students construct and justify their explanations.

Review the following questions with students:

- What is a mixture? [*Answer: A mixture consists of two or more substances, each retaining its own properties.*]

- How did you separate the mixtures? [*Answer: By using a magnet, by sorting, and by sifting.*]

- Were there any other ways to separate them? Were there any mixtures you could not separate?

- What type of change did we make to the materials when we mixed them? [*Answer: Physical.*]

- What type of change did we make to the materials when we separated them? [*Answer: Physical.*]

- Did we have any variables to control in this activity? [*Answer: No.*]

- If there are no variables to control, then we may not be doing an experiment but instead are doing a scientific investigation. Was this an experiment or an investigation? [*Answer: Investigation.*]

Have students make a concluding statement (claim) to the key question. Remind students to use evidence from investigations to support their claims.

ELABORATE

- Describe how students will develop a more sophisticated understanding of the concept.

- What vocabulary will be introduced, and how will it connect to students' observations?

- How is this knowledge applied in our daily lives?

---

1. Ask students to bring in a small sample of their favorite cereal in a baggie. Have them determine if the cereals are mixtures. Let students decide how they would separate any of the cereals that are mixtures.

2. Let students play a "Mixture or Not" game. List 10 items on the board, such as the following: trail mix, fog, salad, Kool-Aid, salt, sugar, roll of Life Savers, pumice, closet full of clothes.

3. Ask students to decide if the items are mixtures or not and explain their decisions. Mixtures include trail mix, Kool-Aid, smoke, fog, salad, Life Savers, closet full of clothes, and pumice. (Fog is a mixture of water and air. Pumice is a mixture of rock and air. Smoke is a mixture of ash and air. Both sugar and salt are compounds. The elements making up these compounds are chemically combined and cannot be separated by ordinary means.)

StEMT*ify*

Science: What real-world issues and problems exist in the area of *human impact on the Earth?*

Engineering: Is the STEM lesson guided by the engineering design process?

- Students should be immersed in hands-on inquiry and open-ended exploration.
- Students should be involved in productive teamwork.

Mathematics: Does the STEM lesson rigorously apply mathematics and science content that the students are learning?

Technology: Does the STEM lesson allow for multiple right answers through prototype development, testing, and improvement of design? Encourage students to monitor and evaluate their progress, change course if necessary, and ask, "Does this make sense?"

Teacher and Student Background

The water Americans get from their faucets is generally safe. This water has been treated and purified. But at least 20% of the world's people do not have clean drinking water. Their only choice may be to drink water straight from a river. If the river is polluted with waste, it will contain bacteria and other organisms that cause disease. Almost nine out of ten cases of disease worldwide are caused by unsafe drinking water. Diseases from unsafe drinking water are the leading cause of death in young children. The causes of pollution are vast. They include such things as agricultural runoff, industry, and the activities of everyday life, including garbage removal, wastewater, building construction, and gas for vehicles.

StEMT*ified* Question: How can we use knowledge of how mixtures can be separated and engineering design practices to design a solution to provide cleaner water to people around the globe?

Step 1: Ask—Practice of asking questions (science) and defining problems (engineering)

What is the problem to solve? What needs to be designed, and who is it for? What is the goal?

In many areas of the world, the water people drink is a mixture of water and other material and organisms. The goal is to create a first-stage filtering system to rid the water of as many pollutants as possible before it would flow into a secondary filtering system to get rid of micro-organisms.

Materials: Small rocks, small hollow beads or Cheerios, plastic tubing, funnel, sand, gravel, potting soil, beakers, aluminum pans, paper or foam cups, clay, straws, small piece of window screen, indirectly vented chemical splash goggles, nonlatex gloves, nonlatex aprons

Step 2: Research and design—Practice of planning and carrying out investigations

What are the project requirements and limitations? You may include as many other materials as you have available for this challenge.

Step 3: Plan—Practice of constructing explanations and designing solutions

1. Students will be given a beaker containing a mixture of water, Cheerios (or any cereal that will float), potting soil, and rocks of varying size.

2. Students will be given time to develop a plan for filtering out each of the materials that are added to the water in a series of steps.

3. Students will design a process to remove each type of material individually, one material removed per step. For example, in Step 1, the material that floats may be removed by skimming the surface of the water. In another step, heavier items may be allowed to sink to the bottom and the cleaner liquid drained off the top.

Step 4: Create—Practice of developing and using models

Students will build a model to demonstrate how mixtures can be separated. Students should remove each type of material individually, one material removed per step. Guiding questions:

(1) How will floating material be removed? (2) How will heavier items that sink be removed? (3) What is left? (4) Is the water left behind clean enough for human consumption?

Step 5: Test and Improve—Practice of obtaining, evaluating, and communicating information

Once it is approved by the teacher, students will construct and test their preliminary design. Students will be given additional time to redesign their apparatus to improve the effectiveness of the design.

Look-Fors: Students should engage in the following practices:

- *Practice of planning and carrying out investigations.* Students will be actively engaged and work cooperatively in small groups to complete investigations, test solutions to problems, and draw conclusions. Use rational and logical thought processes and effective communication skills including writing, speaking, and listening; SEP 7, SEP 8; MP 3).

- *Practice of constructing explanations and designing solutions.* Students will design, plan and carry out investigations to collect and organize data (SEP 3; MP 1).

EVALUATE

- How will students demonstrate that they have achieved the lesson objective?

- Evaluation should be embedded throughout the lesson as well as at the end of the lesson.

Summative Assessment: Write about it!

Once the design has been completed, students will draft a report describing their design. The purpose of this report is to market their product to a Third World community struggling to provide clean water to its people and must include a description of the original mixture, how it was separated, and how this design may change the planet.

- *Claim.* What does the product do?

- *Evidence.* What evidence is there that the product works?

- *Reasoning.* How does this evidence support the claim? How will the design help people?

Reference

National Research Council (NRC). 2012. *A framework for K–12 science education: Practices, crosscutting concepts, and core ideas.* Washington, DC: National Academies Press.

8 | Life Sciences StEMT Lessons

I have always put limits on what I would teach my kids because I was afraid they would not be able to handle very much, but after working with these lessons I realize I was not helping but holding the kids back. These lessons really show how the inquiry skills from the kids are only slowed by what we tell them. If we let them do more, then we will see more.

—Hunter at Lena Vista Elementary

Cell-fie

Life Sciences: From Molecules to Organisms—Structure and Function (LS1.A)

Teacher Brief

What student could resist a lesson that involves a zombie apocalypse? The engineering design challenge of this lesson really ups the excitement level of studying of a cell. The original lesson starts with the typical question: How are animal and plants cells similar and dissimilar? The Explore section has students using microscopes to look at plant cells. The goal of the Explore section is for students to familiarize themselves with the parts of a plant cell. The understanding of how the cell parts work together occurs during the Explain portion of the lesson, as students work in small groups and use the cell structure cards to complete the cell structure and function chart and diagrams. They will also learn how plant and animal cells differ through the use of the cards. This would also be a good place to Explore cheek cell observations. A common follow-up to this lesson is to have students build models of cells, which is certainly appropriate and something you may even consider doing before moving on to the engineering design challenge.

The engineering design challenge for this lesson includes the popular theme of zombies. Teachers should take this opportunity to discuss the difference between science and pseudo science. The zombies in this lesson are merely a fun backdrop for students to use as a system blueprint so students can make the connection that a cell functions as a system, just like a city full of zombies. We encourage you to use your imagination on this lesson. If zombies are not your thing, or have fallen from fashion, there is always the factory, airplane, or some other type of social or mechanical system for comparison. We have tested this lesson in middle school classrooms, and each time the teacher took some liberties with the engineering design challenge portion of the lesson. In all cases, students were actively engaged in the lesson and had fabulous discussions during the process.

Prior Knowledge .

By the end of grade 5. Plants and animals have both internal and external structures that serve various functions in growth, survival, behavior, and reproduction.

By the end of grade 8. All living things are made up of cells, which is the smallest unit that can be said to be alive. An organism may consist of one single cell (unicellular) or many different numbers and types of cells (multicellular). Unicellular organisms (micro organisms), like multicellular organisms, need food, water, a way to dispose of waste, and an environment in which they can live.

Framework Question: How do the structures of organisms enable life's functions?

StEMT*ified* Question: How can we use models to understand phenomena that can be observed at one scale but may not be observable at another scale (e.g., cells as systems)

Dimension 1: Practices

- Developing and using models
- Engaging in argument from evidence
- Obtaining, evaluating, and communicating information

Dimension 2: Crosscutting Concepts

- *Scale, proportion, and quantity.* In considering phenomena, it is critical to recognize what is relevant at different measures of size, time, and energy and to recognize how changes in scale, proportion, or quantity affect a system's structure or performance. Phenomena that can be observed at one scale may not be observable at another scale.

- *Systems and system models.* Defining the system under study—specifying its boundaries and making explicit a model of that system—provides tools for understanding and testing ideas that are applicable throughout science and engineering. Students will develop and use a model to describe how the parts of cells contribute to the function of the system as a whole.

- *Structure and function.* The way in which an object or living thing is shaped and its substructure determine many of its properties and functions. Complex and microscopic structures and systems can be visualized, modeled, and used to describe how their function depends on the relationships among its parts; therefore complex natural structures can be analyzed to determine how they function.

Dimension 3: Disciplinary Core Idea

Structure and function: A central feature of life is that organisms grow, reproduce, and die. They have characteristic structures (anatomy and morphology), functions (molecular-scale processes to organism-level physiology), and behaviors (neurobiology and, for some animal species, psychology). Organisms range in composition from a single cell (unicellular micro-organisms) to multicellular organisms in which different groups of large numbers of cells work together to form systems of tissues and organs (e.g., circulatory, respiratory, nervous, musculoskeletal) that are specialized for particular functions (NRC 2012, p. 143).

Instructional Duration: Three 45-minute class periods

Materials

Per group of three to four students: Cell structure cards (cell wall, cell membrane, nucleus, cytoplasm, chloroplasts, mitochondria, and vacuole; suggested resource: *https://quizlet.com/57013/ cell-structure-flash-cards*), medicine dropper, onion pieces, iodine stain, microscope, forceps, prepared slide or image of an animal cell such as a cheek cell or muscle cell, indirectly vented chemical splash goggles, nonlatex gloves, nonlatex aprons

Safety Notes

1. Personal protective equipment (splash goggles, gloves, and aprons) should be worn during the setup, hands-on, and takedown segments of the activity.

2. Never eat food items used in this activity (risk of cross-contamination).

3. Use caution when working with iodine: It is toxic and can stain clothing and skin.

4. Handle slides and cover slips with caution, as the sharp edges can cut or puncture skin.

5. Wash hands with soap and water upon completing this activity.

For a basic overview of cell structure, including the differences between plant and animal cells, teachers may use the OER CK–12 Standards-Aligned FlexBook textbook *Life Science: Cellular Structure and Function* (*www.ck12.org/teacher*).

Guiding Question: How can models be used to describe the function of a cell as a system as well as part of a system?

Misconceptions (AAAS Project 2061)

- Living things are made of cells; nonliving things are made of atoms.

- Plant cells contain chloroplasts and animal cells contain mitochondria.

- All cells are the same size and shape (i.e., there is a generic cell); some living parts of organisms are not made of cells; plants are not made of cells.

ENGAGE

- How will you capture students' interest and reveal misconceptions?
- What kinds of questions should the students ask themselves after the engagement?

Is it a plant or animal cell? Have students examine several organelles of plant and animal cells (cell wall, cell membrane, nucleus, cytoplasm, chloroplasts, mitochondria, and vacuole). Ask students if they think each image is a plant or animal cell and why (do not provide answers; ask probing questions to guide students in the construction of their own meaning of key terms). See Table 8.1.

Table 8.1. Data collection chart for Engage activity for Lesson LS1.A

Image	Plant or Animal Cell?	Why?
1.		
2.		
3.		

EXPLORE

- Describe the hands-on activities that will provide experience of the phenomenon.

- List "big idea" conceptual questions you will use to encourage and focus students' exploration and to allow the students to test their ideas.

Examining Plant and Animal Cells (onion and cheek cell lab)

1. Get a glass slide and cover slip and make sure they are both thoroughly washed and dried.

2. With the forceps, remove a single layer from the inner (concave) side of the onion piece and place it on the slide. If it is folded or wrinkled, straighten it with the forceps. (Note: It is important to model appropriate use of the microscope and the peeling of a single layer of onion cells from the inner curvature of a piece of onion.)

3. Place a drop of iodine stain on the onion tissue; carefully place a cover slip over the stained onion tissue; gently tap out any air bubbles.

4. Observe the onion cells under low (4×) and medium (10×) power. Draw and label an adjoining group of at least 10 cells on Plate 1. Repeat steps for cheek cell observation (animal cell).

5. Students should record their observations (sketch) for each cell type under magnification.

Ask students the following questions (*Student answers will vary and, at this point, may not include the scientific vocabulary / names of the organelles*):

1. How are plant and animal cells alike?

2. How are plant and animal cells different?

Students should record their observations in their student notebooks.

EXPLAIN

- Student explanations should precede your explanations or introduction of terms. What questions or techniques will you use to help students connect their exploration to the concept under examination?

- List higher-order thinking questions that you will use to solicit *student* explanations and to help students construct and justify their explanations.

Plant and Animal Cells as a System

Note: *Do not answer questions.* Listen for misconceptions and asks further probing questions in order to guide the discussion. Results lead into Explain phase of 5E.

1. Distribute *cell structure cards* (cell wall, cell membrane, nucleus, cytoplasm, chloroplasts, mitochondria, and vacuole), one set per group.

2. Model with the card labeled "Cell Membrane" to show students how the table is to be completed.

3. In small groups, students use the cards to complete the cell structure and function chart and diagrams. (See Table 8.2) Ask students the following questions, ensuring that students use scientific vocabulary with regard to the cell parts. Student answers from the Explore section should be modified as needed based on new information from the Explain section.

- How are plant and animal cells alike? [*Answer: Plant and animal cells both contain a cell membrane, cytoplasm, a nucleus, vacuoles, and mitochondria.*]

- How are plant and animal cells different? [*Answer: Plant cells contain chloroplasts and a cell wall and animal cells do not. Plant cells also have a large central vacuole.*]

Refer back to the cell structure cards with images of the cell wall, cell membrane, nucleus, cytoplasm, chloroplasts, mitochondria, and vacuole, and ask students again if each of the images is representative of a plant or animal cell. Sketch and label the identified cell parts.

Table 8.2. Data collection chart for Explain activity for Lesson LS1.A

Criteria	Cell Membrane	Cell Wall	Nucleus	Cytoplasm	Chloroplasts	Mitochondria	Vacuoles
Description	thin layer of protein/fat around a cell						
Function	controls what enters/leaves cell						
Plant Cell	Yes						
Animal Cell	Yes						

ELABORATE

- Describe how students will develop a more sophisticated understanding of the concept.

- What vocabulary will be introduced, and how will it connect to students' observations?

- How is this knowledge applied in our daily lives?

When students understand the basic differences between a plant and animal cell, assign pairs to build a model of an animal cell, choosing materials from a variety of items that you provide. Student pairs briefly discuss the types of items they could use to represent the cell structures identified in the previous activity. Students should gather materials (two plastic sandwich baggies per student pair to represent the cell membrane; a variety of materials to represent cell parts, such as buttons, pasta of different colors, pipe cleaners, and beads; one cup of clear syrup, or similar, for each student pair) and make the cells. Provide students with goggles, nonlatex gloves, and nonlatex aprons. The chart in Table 8.3 (p. 84) should be completed in the students' notebooks.

Table 8.3. Data collection chart for Elaborate activity for Lesson LS1.A

Structure	Plant	Animal	Function	Material Used	Justify Why Material Used
Cell wall					
Cell membrane					
Nucleus					
Cytoplasm					
Chloroplasts					
Mitochondria					
Vacuole					

Safety Notes

1. Personal protective equipment (goggles, gloves, and aprons) should be worn during the setup, hands-on, and takedown segments of the activity.

2. Never eat food items used in this activity (risk of cross-contamination).

3. Use caution when working with sharps (pipe cleaners) that can cut or puncture skin.

4. Wash hands with soap and water upon completing this activity.

Students should put the items representing the various cell parts into the baggies before they pour in the syrup so that they can promptly seal the bag once the syrup is poured. Once the "cell structures" are in the baggie, have students add the syrup. Have them pour the syrup into a measuring cup that has a spout for easy pouring. One student should carefully hold the baggie with both hands as the other pours in the syrup. While making their models, students should continue working on the data chart and record the function of each structure. In addition, students should record the material they chose to represent each cell structure as well as the reason for doing so (i.e., indicate how the material is representative of the particular structure). After students have made their model cells, allow students to compare their models and discuss the similarities and differences.

Ask the students the following questions:

- Why do we often depend on models? Why are models useful when discussing cells?

- How is your model like a real cell? How is it different?

- What are some limitations of models in general?

- How does this model represent a system?

StEMT*ify*

Science: What real-world issues and problems exist in the area of system design?

Engineering: Is the STEM lesson guided by the engineering design process?

- Students should be immersed in hands-on inquiry and open-ended exploration.

- Students should be involved in productive teamwork.

Mathematics: Does the STEM lesson rigorously apply mathematics and science content that the students are learning?

Technology: Does the STEM lesson allow for multiple right answers through prototype development, testing, and improvement of design? Encourage students to monitor and evaluate their progress, change course if necessary, and ask, "Does this make sense?"

StEMT*ified* Question: How can we use models to understand phenomena that can be observed at one scale but may not be observable at another scale?

Students engage in the analysis of parts, subsystems, interactions, and matching. The descriptions of parts and their interactions are more important than just calling everything a system (AAAS 1993, p. 265). In addition, studies of student thinking indicate that they tend to interpret phenomena by noting the qualities of separate objects rather than by seeing the interactions between the parts of a system (AAAS 1993, p. 355). In the context of cells, encourage students to look at the cell as both a system and a subsystem and to develop an understanding of how the parts of a cell interact with one another. Ask students the following questions:

- With what systems are you familiar? [*Examples: school system, solar system, digestive system, pulley system.*]

- What makes these things systems? How would you define a system?

- Is a city a system? Why or why not? Is a factory a system? Why or why not?

- What other things would you consider systems? Explain why.

Post–Zombie Apocalypse City (or factory or airplane if zombies are not trending)

Students will design a self-sustaining post–zombie apocalypse city and consider how a city and a cell can both be thought of as systems. This is a teachable moment to discuss science (cell theory) versus pseudoscience (zombie apocalypse). Divide the class into small groups of two or three.

Step 1: Ask—Practice of asking questions (science) and defining problems (engineering)

What is the problem to solve? What needs to be designed, and who is it for? What are the project requirements and limitations? What is the goal?

Step 2: Research and design—Practice of planning and carrying out investigations

Students collaborate to brainstorm ideas and develop as many solutions as possible. How are the parts of a city, factory, or airplane similar to the structures of organisms? How do the parts interact to allow the city, factory, or airplane to function as a unit?

Step 3: Plan—Practice of constructing explanations and designing solutions

Students will compare the best ideas, select one solution and make a plan to investigate it. Students should complete the charts in Table 8.4 on page 87 and Table 8.5 on page 88. Students may refer to various readings about Cell Factory, Cell City, OER CK-12 Cell Structure and Function, etc., at discretion of teacher) to complete the charts. (**Required**: cytoplasm, nucleus, cell/plasma membrane, cytoskeleton/cell wall, mitochondria, and chloroplasts. **Extended Learning**: endoplasmic reticulum, ribosomes, Golgi apparatus, lysosomes).

EVALUATE

- How will students demonstrate that they have achieved the lesson objective?

- Evaluation should be embedded throughout the lesson as well as at the end of the lesson.

Table 8.4. Data collection chart for StEMT activity for LS1.A.

Comparing a Cell to City/Factory/Airplane		
Job in Zombie-Free City (Factory) (Airplane)	**Cell Organelle**	**Function of the Organelle**
City Border Patrol (*Shipping/receiving department*) (Cockpit)	*Plasma/cell membrane*	*Regulates what enters and leaves the cell; protects cell*
City Hall (*Chief Executive Officer*) (Pilot)	*Nucleus*	*Controls all major cell activity; determines what proteins are made*
Road System (*Factory floor*) (Air in the cabin)	*Cytoplasm*	*Contains the organelles; site of most cell activity*
Industries (*Assembly line*)	*Endoplasmic Reticulum (ER)*	*Is where ribosomes do their work*
Farmers and Ranchers (*Workers in the assembly line*)	*Ribosomes*	*Build the proteins*
Packaging Companies (*Finishing/packaging department*)	*Golgi apparatus*	*Prepares proteins for use or export*
Recycling/Garbage company (*Maintenance crew*)	*Lysosomes*	*Responsible for breaking down and absorbing materials taken in by the cell*
City Limits (*Support: walls, ceilings, floors*) (Steel beams and frame of the hull)	*Cytoskeleton/ Cell wall*	*Maintains cell shape*
Power Company (*Power plant*) (Jet engines and jet fuel)	*Mitochondria / chloroplasts*	*Transforms one form of energy into another*
Water Plant (*Water plant*) (Overhead storage compartments)	*Vacuole*	*Water storage*

Step 4: Create — Practice of developing and using models

Students will build a prototype of a city and label the parts of the system using city terms and cell terms.

Step 5: Test and Improve — Practice of obtaining, evaluating, and communicating information

Does the prototype meet the goal and fall within the limitations and requirements established in Step 1? Students communicate the results and get feedback, and analyze and talk about what works, what does not and what could be improved.

Table 8.5. Data collection chart for Evaluate activity for Lesson LS1.A

Question	Plant/Animal Cell	City/Factory	Analogy (Plane, School)
When this system is working, what does it do?	Produces proteins		
For this system to work, must it receive any input?	Energy from the Sun		
What, if any, output does this system produce?	Proteins		
Identify at least two parts of this system that must interact if the system is to function. Describe how these parts interact.			
Identify any subsystems within the whole system.			

Claims, Evidence, Reasoning

How are the parts of a city, factory, or airplane similar to the structures of organisms? How do the parts interact to allow the city, factory, or airplane to function as a system? Use evidence from the model to describe the parts and their interactions. What would happen to the city, factory, or airplane if one of the parts were removed? Explain your reasoning.

Summative Writing Assignment (see sample writing rubric in Appendix A)

In approximately 300 words, create and explain an analogy to complete the sentence *A cell is like a [plane, school, etc.]* (to differentiate, ELL students may write about the analogy done in class but students should be given the opportunity to create a new analogy). Connect the functions of the organelles studied in class to the new analogy. The essay will explain how the analogy *works*. Identify at least four parts of this system. Describe what each part does, and tell how each part contributes to the system as a whole. Include the following:

- Can any one part of the system do what the whole system does? Justify your response. (Answers will vary. Students should realize that the organelles need to work together to produce proteins.)

- Describe how the functioning of this system would change if one of the parts wears out.

- In what ways is it useful to think of the cell as a system?

Sample Writing Rubric for Cell Structure and Function

Analogy. Writer presents a unified analogy comparing the structure of a cell to another self-contained system. The comparison incorporates and connects all the cell's different organelles into the explanation of the analogy. The choice of analogy is novel and demonstrates the writer's understanding of cell structure.

Essay Structure. Essay is organized by a strong introduction that establishes the analogy, followed by a body where the analogy is elaborated on and explained, and ending with a conclusion. The paragraphs are connected by transitions between the beginning, middle, and end. Organelle names are highlighted.

Explanation. Tone of the essay is informational. Writer uses the analogy to clearly explain how a cell works and uses the organelles to help make the point. Writer does not lose the science of cell structure in empty adjectives. Words are well chosen and work to elaborate on the analogy and cell structure. It is clear that the writer knows what she or he is talking about.

Itsy, Bitsy Spiders

Ecosystems: Interactions, Energy, and Dynamics—
Interdependent Relationships in Ecosystems (LS2.A)

Teacher Brief

With climate change as front-page news, this lesson seems very appropriate to include in your arsenal. If you are doing this lesson with younger students, we recommend avoiding the political implications, but with older students it might be included as a great beginning hook to this lesson. The Explore part of many lessons on this topic has students doing research, which can be very empowering to students as it gives them the opportunity to be the expert. We have provided just a basic format for the share-out of student research, but there are many creative ways to provide other options for this portion. There are also simulations such as the PHET Simulation on Natural Selection from the University of Colorado that can be used to demonstrate how animals of different colors are affected by changes in their environment.

The engineering design challenge we have included is a lot of fun for students and a great opportunity for a teacher to visit each group and discuss the guiding questions as students create their design. With all engineering design challenges, it is important to allow students time to develop their designs by testing and making changes based on those tests. We have purposefully tried to not include an exhaustive list of material for this and many of the other engineering challenges. We have found that students are far more creative when they can choose or ask for materials. One way to do this is to have groups make a list of potential items the day before the challenge, giving the teacher time to collect the requested materials.

Prior Knowledge

By the end of grade 5. The food of almost any kind of animal can be traced back to plants. Organisms are related in food webs in which some animals eat plants for food and other animals eat the animals that eat plants. Either way, animals are "consumers." Organisms can survive only in environments in which their particular needs are met. A healthy ecosystem is one in which multiple species of different types are each able to meet their needs in a relatively stable web of life. Newly introduced species can damage the balance of an ecosystem (*Framework;* NRC 2012, p.152).

By the end of grade 8. Organisms and populations of organisms are dependent on their environmental interactions, both with other living things and with nonliving factors. The growth of organisms and population increases are limited by access to resources. In any ecosystem, organisms and populations with similar requirements for food, water, oxygen, or other resources may compete with each other for limited resources, access to which consequently constrains their growth and reproduction. Similarly, predatory interactions may reduce the number of organisms or eliminate whole populations of organisms. Mutually beneficial interactions, in contrast, may cause organisms to become so interdependent that each organism requires the other for survival. Although the species involved in these competitive, predatory, and mutually beneficial interactions vary across ecosystems, the patterns of interactions of organisms with their environments, both living and nonliving, are shared (*Framework;* NRC 2012, p. 152).

Framework Question: How do organisms interact with the living and nonliving environments to obtain matter and energy (LS2.A)?

StEMT*ified* Question: How can a web be designed to catch many types of insects?

Dimension 1: Practices (What should students be doing?)

- Developing and using models
- Analyzing and interpreting data
- Engaging in argument from evidence
- Obtaining, evaluating, and communicating information

Dimension 2: Crosscutting Concepts

- *Cause and effect: Mechanism and explanation.* Events have causes—sometimes simple, sometimes multifaceted. A major activity of science is investigating and explaining causal relationships and the mechanisms by which they are mediated. Such mechanisms can then be tested across given contexts and used to predict and explain events in new contexts.

- *Energy and matter: Flows, cycles, and conservation.* Tracking fluxes of energy and matter into, out of, and within systems helps one understand the systems' possibilities and limitations. Matter cycles between the air and soil and among plants, animals, and microbes as these organisms live and die.

• *Systems and system models.* Defining the system under study—specifying its boundaries and making explicit a model of that system—provides tools for understanding and testing ideas that are applicable throughout science and engineering. A system can be described in terms of its components and their interactions.

Dimension 3: Disciplinary Core Idea

Ecosystems are ever changing because of the interdependence of organisms of the same or different species and the nonliving (physical) elements of the environment. Within any one ecosystem, the biotic interactions between organisms (e.g., competition, predation, and various types of facilitation, such as pollination) further influence their growth, survival, and reproduction, both individually and in terms of their populations (*Framework;* NRC 2012, p. 151). Sometimes the differences in characteristics between individuals of the same species provide advantages in surviving, finding mates, and reproducing (LS4).

Instructional Duration: Four 45-minute class periods

Materials per Class of 30

Pictures of white Arctic fox, brown Arctic fox, snowshoe hare (projected or in sets); map of polar climate zone (see StEMT materials list in Elaborate section)

For a basic overview of adaptations and camouflage, teachers may use the OER CK-12 Standards-Aligned FlexBook textbook *Life Science:Ecology* at *www.ck12.org/life-science/ Consumers-and-Decomposers-in-Life-Science/lesson*

Guiding Questions

• In what ways do organisms adapt to changes in climate (biotic and abiotic factors)?

• How do changes in climate lead organisms to die or move to new locations?

Misconceptions

Food is a source of building materials but not a source of energy, and animals cannot store molecules from food in their bodies (AAAS Project 2061, n.d).

Students write the key question in their science notebook and brainstorm any preliminary thoughts for discussion with a partner or their group. Circulate the room to identify misconceptions and determine preliminary knowledge. Show students the pictures of the Arctic fox in the two phases (winter and summer camouflage).

Ask probing questions for small-group discussion:

- What do these pictures show? Is this the same animal or two different animals? Why would this fox be able to look two different ways? Why would a summer coat and a winter coat be a beneficial adaptation?

- How do other animals adapt to changes in climate?

- Why would a black bear have difficulty surviving in the arctic or polar climate zone?

- If the polar zone has the most extreme variation in temperatures, how would that affect their adaptations to survive?

- How do biotic and abiotic factors play a role in organism survival?

Prior Knowledge
Teacher Notes

Many arctic animals have extreme adaptations to their environment. Arctic foxes, snowshoe hares, and snowy owls have a completely different coat color in winter than in summer and look like totally different animals. For example, the snowshoe hare's coat changes from brown to white and back again. When the snow begins to fall in autumn or early winter, the hare begins to grow a new, lighter coat. By the time winter arrives, the coat will be completely white. This color change helps the hare hide from predators by blending into the white snow. In the spring, when the snow melts, the snowshoe hare grows another brown coat. Elementary students should not be exposed to the political issues connected to the idea of climate change. Students can still explore what is happening in the polar region without having to discuss the varying views on climate change. It is common knowledge that many of the ice caps are melting and that the environment in the polar areas is changing. Explain to students that they are going to illustrate how this adaptation can be beneficial.

Pair students to do research on polar climates. Each student creates a habitat poster that includes the foliage, precipitation, and other elements of the habitat in either winter or summer. The pairs will end up with one winter poster and one summer poster. Once posters are done, call the pairs together.

- Ask students: What did you learn about the polar climate? Why do you think the Arctic foxes changing coat helps it survive?

- Tell students to place the white rabbit die cut on the winter poster.

- Ask students: Do you think that this organism's camouflage, which is an adaptation, would help it survive in this environment? Why?

- Tell students to place the white rabbit die cut on the summer poster.

- Ask students: Do you think that this organism's camouflage, which is an adaptation, would help it survive in this environment? Why?

- Tells student pairs to do the same with the other die cuts and to discuss how these camouflage adaptations would help or hurt the animal.

EXPLAIN

- Student explanations should precede your explanations or introduction of terms. What questions or techniques will you use to help students connect their exploration to the concept under examination?

- List higher-order thinking questions that you will use to solicit *student* explanations and to help students construct and justify their explanations.

Ask students:

- Which adaptation worked for the rabbits in each season?

- Why? Similarity: Like the arctic fox, the snowshoe hare, or arctic rabbit, also changes coat color depending on the season.

Ask students:

- Which adaptation worked for the owls in each season? Why? Similarity: Like the arctic fox and the snowshoe hare, the snowy owl also changes coat color depending on the season.

Ask students:

- Which adaptation worked for the bears worked in each season? Why? Difference: Bears do not change coat color.

Ask students:

- If you were a black bear, native to Florida, would you survive in the Arctic?
- If an organism cannot adapt and the entire species dies out, what is that called? Ask for examples of species that have gone extinct (died out).
- In what ways do organisms adapt to changes in climate?
- How do changes in climate lead organisms to die or move to new locations?

Have students write an updated response to the key question. Remind them to use evidence from the activities or from research as support for their claims.

ELABORATE

- Describe how students will develop a more sophisticated understanding of the concept.
- What vocabulary will be introduced, and how will it connect to students' observations?
- How is this knowledge applied in our daily lives?

Have students create an illustration of an arctic food web using ThinkQuest (*library. thinkquest.org/3500/animals.htm.*) and have students research the arctic environment and the changes that are happening to the environment. In their research, have students examine the reasons that some of the organisms, such as polar bears, are dying out due to climate change. Teach the AIMS lesson "Missing Moths" to extend the instruction on adaptation to environmental change. This resource is available at the following web site and is aligned to the SEP Engaging in Argument from Evidence: *http://ngss.nsta.org/Resource.aspx?ResourceID=660.*

> ## StEMT*ify*
>
> **S**cience: What real-world issues and problems exist in the area of *ecosystem interactions?*
>
> **E**ngineering: Is the STEM lesson guided by the engineering design process?
>
> - Students should be immersed in hands-on inquiry and open-ended exploration.
> - Students should be involved in productive teamwork.
>
> **M**athematics: Does the STEM lesson rigorously apply mathematics and science content that the students are learning?
>
> **T**echnology: Does the STEM lesson allow for multiple right answers through prototype development, testing, and improvement of design? Encourage students to monitor and evaluate their progress, change course if necessary, and ask, "Does this make sense?"

Step 1: Ask—Practice of asking questions (science) and defining problems (engineering)

What is the problem to solve? What needs to be designed and who is it for?

The climate of a region has changed, causing some insects to relocate or not survive. The spiders of this region cannot relocate. Spiders catch their food by constructing webs. As their typical food sources are no longer available, they may be challenged to catch new types of insects to replace the loss of their normal diet.

StEMT*ified* Question: How can a web be designed to catch as many types of insects as possible?

What are the project requirements and limitations?

Materials: Ten meters of yarn; one meter of masking tape; Styrofoam packing peanuts, safety glasses or goggles

Safety Notes

1. Personal protective equipment (safety glasses or goggles) should be worn during the setup, hands-on, and takedown segments of the activity.

2. Use caution working with yarn (potential trip-and-fall hazard).

3. Wash hands with soap and water upon completing this activity.

Step 2: Research and design—Practice of planning and carrying out investigations

Students collaborate to brainstorm ideas and develop as many solutions as possible.

1. Place two desks 30 cm apart.

2. Pass out materials to each group. Students may only use the materials provided.

3. Allow students to build a web that will catch a Styrofoam packing peanut falling from a height of one meter. This step simulates the spider's normal catch.

Step 3: Plan—Practice of constructing explanations and designing solutions

Students will compare the best ideas, select one solution, and make a plan to investigate it. Explain that the environment has changed, causing many of the spiders' prey to either move or die. To survive, the spider must be able to catch many different types of insects. Present objects of a slighter heavier mass and smaller in size to drop. Show students the objects.

Step 4: Create—Practice of developing and using models

Students will build a prototype. Students will design and redesign their web in order to catch the object. Explain that if their web does not catch the object that represents the spiders new prey, the spider they represent may not survive.

EVALUATE

- How will students demonstrate that they have achieved the lesson objective?

- Evaluation should be embedded throughout the lesson as well as at the end of the lesson.

Step 5: Test and improve—Practice of obtaining, evaluating, and communicating information

Does the prototype work and does it solve the need? Students communicate the results and get feedback. Students then talk about what works, what does not work, and what could be improved.

Summative Assessment: Write about it!

Use evidence to construct an explanation (claim) for the best web design to catch many types of insects. How do the variations in characteristics among individuals of the same species provide advantages in surviving? Use data collected from Explore and StEMT*ify* sections (web design) to support your claim.

Gobble, Gobble, Toil and Trouble

Life Sciences: Ecosystems: Interactions, Energy, and Dynamics—Ecosystem Dynamics, Functioning, and Resilience (LS2.C)

Teacher Brief

A teacher wrote us a note stating that this was the first time all year she had 100% engagement in a lesson. The Turkey Trot lesson has become very popular in the middle school where we cover limiting (biotic and abiotic) factors in environments. Using a game, for Explore students will interact with the limiting factors on a turkey population, which is essentially a collection of data that can be referenced during the Explore debrief. This is also a great opportunity for students to think about mathematical modeling using their population data. A traditional extension to this lesson would be to use text that brings in a regional or local example of invasive species.

The engineering design challenge we have included involves designing and building a scale model of a new levee system needed to protect a city against the high water associated with a hurricane. To give this activity a more local feel, this storyline could easily be adapted for flooding due to heavy rains or a high rate of snowmelt. In this lesson, we tried to emphasize research and planning as students develop their ideas for their prototype. Make this part as elaborate as you want, with time being your only limitation. We have included ideas of how to bring in social and economic considerations as well.

Prior Knowledge

By the end of grade 5. When the environment changes in ways that affect a place's physical characteristics, temperature, or availability of resources, some organisms survive and reproduce, others move to new locations, others move into the transformed environment, and some die (*Framework;* NRC 2012, p. 155).

By the end of grade 8. Ecosystems are dynamic in nature; their characteristics can vary over time. Disruptions to any physical or biological component of an ecosystem can lead to shifts in all of its populations. Biodiversity describes the variety of species found in Earth's terrestrial and oceanic ecosystems. The completeness or integrity of an ecosystem's biodiversity is often used as a measure of its health (*Framework;* NRC 2012, p. 155).

Dimension 1: Practices (What should students be doing?)

- Developing and using models (using a simulation or model to demonstrate real-life events in the form of a game for analysis and prediction)

- Analyzing and interpreting data (recognizing patterns and making inferences about changes in populations; small changes in one part of a system might cause large changes in another part of a system)

- Using mathematics, information and computer technology, and computational thinking

- Engaging in argument from evidence

Dimension 2: Crosscutting Concepts

- *Cause and effect: Mechanism and explanation.* Events have causes—sometimes simple, sometimes multifaceted. A major activity of science is investigating and explaining causal relationships and the mechanisms by which they are mediated. Such mechanisms can then be tested across given contexts and used to predict and explain events in new contexts.

- *Stability and change.* For natural and built systems alike, conditions of stability and determinants of rates of change or evolution of a system are critical elements of study. Small changes in one part of a system might cause large changes in another part.

Dimension 3: Disciplinary Core Idea

Ecosystems are dynamic in nature; their characteristics fluctuate over time depending on changes in the environment and in the populations of various species. Disruptions in the physical and biological components of an ecosystem—which can lead to shifts in the types and numbers of the ecosystem's organisms, to the maintenance or the extinction of species, to the migration of species into or out of the region, or to the formation of new species (speciation)—occur for a variety of natural reasons. Changes may derive from the fall of canopy trees in a forest, for example, or from cataclysmic events, such as volcanic eruptions. But many changes are induced by human activity, such as resource extraction, adverse land-use patterns, pollution, introduction of non-native species, and global climate change. Extinction of species or evolution of new species may occur in response to significant ecosystem disruption. Species in an environment develop behavioral and physiological patterns that facilitate their survival under the prevailing conditions, but these patterns may be maladapted when conditions change or new species are introduced. Ecosystems with a wide variety of species—that is, greater biodiversity—tend to be more resilient to change than those with few species (*Framework;* NRC 2012, p.155).

Framework Question: What happens to ecosystems when the environment changes? (LS2.C)?

StEMT*ified* Question: How can teams of engineers (students) design a sturdy barrier to prevent water from flooding a city in the event of a hurricane and ensure habitat preservation?

Instructional Duration: Approximately five 50-minute class periods

For a basic overview of speciation, biodiversity, and adaptations, teachers may use the OER CK-12 Standards-Aligned FlexBook Textbook *Life Science* (*https://www.ck12.org/teacher*).

Materials for Turkey Trouble Activity (per group)

- Two Dice
- Turkey Trouble Game Chart
- Article for Close Read (topic should focus on exotics introduced into the wild)

Materials for Levee Activity (per group)

- Rectangular plastic container or tub
- Scissors, to poke holes in cups
- $10 worth of "fun" money (e.g. Monopoly money)
- Levee building materials: sand or gravel (about 2 cups) and duct tape (about 1 foot)
- Eight to ten cotton balls
- One plastic sandwich bag or square of plastic wrap
- One sponge

- Eight to ten craft sticks

- Eight to ten plastic drinking straws

- Paperboard (1 sheet, such as from a cereal box)

- Small paper cups

- Access to water to test the levees

- Indirectly vented chemical splash goggles

- Nonlatex gloves

- Nonlatex aprons

Safety Notes

1. Personal protective equipment (splash goggles, gloves, and aprons) should be worn during the setup, hands-on, and takedown segments of the activity.

2. Wash hands with soap and water upon completing this activity.

Guiding Question: How do limiting factors affect population sizes in a local ecosystem?

Describe how particular ecosystem factors might influence each other (e.g., a shortage of space or nesting sites might increase the likelihood of disease and parasitism).

Misconceptions (AAAS Project 2061, n.d.)

- Varying the population size of a species may not affect an ecosystem because some organisms are not important.

- Species coexist in ecosystems because of their compatible needs and behaviors; they need to get along.

ENGAGE

- How will you capture students' interest and reveal misconceptions?

- What kinds of questions should the students ask themselves after the engagement?

The *Habitat Change* assessment probe (Keeley 2007) is written in the format of a prediction probe, asking students to predict what would happen to divos (imaginary animals living on an island with warm climate and plenty of tree ants, the divo's source of food) if the environment on that island changes dramatically, and all of the tree ants die. Students are asked to "circle any of the things you think happened to most of the divos living on the island after their habitat changed" (Keeley 2007, p. 143). Each of the distractors represents a commonly held student misconception about adaptation; one of the choices represents the scientific conception. The second part of the assessment probe asks students to explain their thinking (i.e., explain how they decided what effect the change in the divos' habitat would have on most of the divos). This formative assessment probe elicits whether or not students think that individuals can intentionally change their physical characteristics (fur length and thickness, or teeth or mouth-parts) or their inherited behaviors (digging holes to live in, hibernating in cold weather) in response to a change in the environment. Share out answers (or four corners activity). Limit feedback; paraphrase for purpose of clarifying.

EXPLORE

- Describe the hands-on activities that will provide experience of the phenomenon.

- List "big idea" conceptual questions you will use to encourage and focus students' exploration and to allow the students to test their ideas.

Turkey Trouble

Students will play the role of state wildlife biologists after they introduce a new animal to the area. The students will simulate environmental changes that affect how the population of turkeys will change over time. They will note the things that cause a decrease in the population and learn that these factors are called limiting factors. For this activity, print the Turkey Trouble game chart (Table 8.6).

Ask the students: Have you ever seen a wild turkey? Have you or a family member been hunting for turkey? Today you are going to assume the role of a wildlife biologist who is studying this type of turkey as we play a game called Turkey Trouble.

Procedure

You are working for the state Department of Natural Resources (DNR) as a wildlife biologist. The DNR wants to introduce a population of Osceola/Merriam turkeys to the panhandle in a wilderness preserve to see if they can become established there. In this activity, you will

Table 8.6. Game chart for Explore activity for Lesson LS2.C

Dice Value	Event	Effect on Population
2	Housing development	People need houses, you know. So sorry, but your entire flock was just eliminated except for one pair that was taken to the local zoo. Unfortunately, both birds were males.
3	Prairie fire	The fire consumed everything! Alas, only 10 birds survived (no matter how many you had to start the round with).
4	Mild winter	Your flock just grew by 75 (no matter how many you had to start the round with).
5	Hunting season	If your flock is less than 200, you are safe because no hunters will drive to your area for so few birds. If your flock is more than 200, then 70 are killed by hunters. If your flock is exactly 200, then the population remains unchanged.
6	Good hatching weather	The size of your flock triples.
7	Disease	If your flock is less than 300, you only lose 20 to the disease. If your flock is larger than 300, you lose 95 of your birds. If your flock is exactly 300, then your flock remains unchanged.
8	Abundant food	Your flock will double if it is currently under 500 birds. It will grow by only 100 if it was larger than 500 to start the round.
9	Corn planted nearby	Your population will triple if it started the round under 99. If your flock is 99 or more, the farmer will be concerned the birds will eat too much corn and will convince the Game & Fish Department to have a hunting season. Then your population will fall by 60.
10	Coyotes move in	If your flock is less than 99, only 20 fall prey to the coyotes since your birds are too scattered to provide a daily food supply for the coyotes. If your flock is 99 or more, then 50 will be eaten by the carnivorous animals.
11	Mild summer	Your flock doubles.
12	Flash flood	All but 10 birds die.

Source: Turkey Trouble, *https://apbiokorzwiki.wikispaces.com/file/view/Turkey+Trouble+AP+Version%5B1%5D.pdf.*

simulate environmental changes that can affect populations of turkeys and will track how the population will change over time as a result of the environmental changes.

- Work in pairs. Construct an appropriate data chart that reflects the data you will collect during each round. Each player must keep track of the number rolled, the event, and the change in the population.

- Each group begins with 100 turkeys. Roll the two dice. Depending on the total points of the toss, use the game chart to determine what you are to do. Notice that there are differences in outcomes depending on the population size.

- Each group will roll the dice at least 20 rounds (unless they are wiped out). Be sure to record information for your population after each round.

- Create a graph of your population as it changed over each of the rounds. Each round represents one year. (What type of graph is appropriate when graphing how things change over time?)

Questions for small-group discussion:
A *limiting factor* is a factor that can limit the growth of a population.

1. What biotic factors affected your population of turkeys? What abiotic factors affected your population of turkeys?

2. Which set of factors (biotic or abiotic) had a greater effect on your population's size and why?

3. Based on your graph, what pattern of population growth do you notice about your turkey population?

4. Which limiting factor has the greatest impact on the turkey population? Justify your answer.

5. What are some other limiting factors that were not used in this game that could limit the growth of the turkey population?

6. How do you think this activity accurately models the increase and decrease of real populations in nature? How could this model be improved to more accurately show what happens to real populations in nature?

All populations of organisms have some effect on their ecosystem, whether positive or negative. Describe how the turkey population affects its local ecosystem, including any interactions it may have with other populations.

EXPLAIN

- Student explanations should precede your explanations or introduction of terms. What questions or techniques will you use to help students connect their exploration to the concept under examination?

- List higher-order thinking questions that you will use to solicit *student* explanations and to help students construct and justify their explanations.

Think–Pair–Share

Explain that the students have tracked the impact on the limiting factors on the population of turkeys. Students write a potential definition for "limiting factors."

Questions

1. What were some events that caused the population of turkeys to decrease?

   ○ List these on the board.

   ○ Introduce terms *biotic* and *abiotic.*

   ○ Have students work in pairs (or small groups to differentiate) to divide the events into biotic and abiotic.

2. What could be some other factors that were not used in this game that could limit the growth of the turkey population?

3. Which limiting factor has the greatest impact on the turkey population? Why?

4. What methods would a wildlife biologist use to study a population of turkeys in the wild?

5. How do factors that limit carrying capacity affect the health, numbers, and distribution of animals?

   ○ How might we determine the carrying capacity of a bucket (basket, etc.)?

   ○ How might we apply this analogy of the capacity of the bucket to the carrying capacity of an ecosystem?

   ○ Write a definition for *carrying capacity* as it applies to an ecosystem.

   ○ Share out, provide feedback, and encourage students to improve definitions.

6. How do these factors affect competition within a species?

7. Why is a balanced habitat important for animals?

8. Are wildlife populations static, or do they tend to fluctuate as part of an overall "balance" of nature? Is nature ever really in "balance," or are ecological systems involved in a process of constant change?

9. Who controls the population fluctuations?

10. What happens when a predator (or prey) is introduced?

Look-Fors

Students should engage in the following practices:

- *Practice of asking questions (science) and defining problems (engineering).* Students will ask questions, define problems, and predict solutions/results (SEP 1; MP 1).

- *Practice of developing and using models.* Students will obtain, evaluate, and communicate information by constructing explanations and designing solutions (SEP 8; MP 3). Students will develop and use models (different types of levee structures) (SEP 2; MP 4).

ELABORATE

- Describe how students will develop a more sophisticated understanding of the concept.

- What vocabulary will be introduced and how will it connect to students' observations?

- How is this knowledge applied in our daily lives?

Non-native species are often introduced into an environment where they compete against the native species that already live there.

Close Read

(Teacher-selected article about local exotics population)

1. Using text features (titles, headings, illustrations, charts, etc.), what might the article be about?

2. Read the article with a partner and underline information that describes the habitat conditions needed for the exotic organism to survive.

3. On sticky notes (or text code with *L* for limiting), write possible limiting factors for the exotic organism. Circle factors (or text code with *N*) that do **not** limit the number of exotics.

4. Include specific questions about the exotic population and limiting factors in the selected article.

StEMT*ify*

Science: What real-world issues and problems exist in the area of wildlife management?

Engineering: Is the STEM lesson guided by the engineering design process?

- Students should be immersed in hands-on inquiry and open-ended exploration.
- Students should be involved in productive teamwork.

Mathematics: Does the STEM lesson rigorously apply mathematics and science content that the students are learning?

Technology: Does the STEM lesson allow for multiple right answers through prototype development, testing, and improvement of design? Encourage students to monitor and evaluate their progress, change course if necessary, and ask, "Does this make sense?"

Look-Fors

Students should engage in the following practices/behaviors:

- *Practice of asking questions (science) and defining problems (engineering).* Students will ask questions, define problems, and predict solutions/results (SEP 1; MP 1).

- *Practice of planning and carrying out investigations.* Students will be actively engaged and work cooperatively in small groups to complete investigations, test solutions to problems, and draw conclusions. Use rational and logical thought processes, and effective communication skills (writing, speaking and listening) (SEP 7, SEP 8; MP 3).

- *Practice of constructing explanations and designing solutions.* Students will design, plan and carry out investigations to collect and organize data (SEP 3; MP 1).

- *Practice of obtaining, evaluating, and communicating information.* Students will analyze and interpret data to draw conclusions and apply understandings to new situations (SEP 4; MP 5). Apply scientific vocabulary after exploring a scientific concept (SEP 6; MP 7).

Protecting Our City With Levees (adapted from Teach Engineering)

StEMT*ified* Question: Acting as engineers for their city, how can teams design a sturdy barrier to prevent water from flooding a city in the event of a hurricane and ensure habitat preservation?

Step 1: Ask—Practice of asking questions (science) and defining problems (engineering)

The flooding from the annual overflow of the Big River every spring has recently been very devastating. The county has the necessary funds to address the issue by building a levee near the section of the Big River where flooding occurs most often. The zoning board has come up with a set of possibilities for the levee design and will hire a team to make recommendations on the final design based on the data to determine the most cost-effective material to use. In addition to the cost of materials, factors such as yearly maintenance, time to complete the levee, and how disruptive the construction will be to the local community must be considered in the team's decision. In order to help expedite this process, the local government is asking teams to design a proposal for their recommendation, which must include the rankings for each type of level (1 = best, 4 = worst) and the process by which the team determined the rankings.

Build a new levee system that will maintain the boundary between the lake or river and the city. A real levee must be five meters higher than sea level and wide enough to prevent the surrounding lake, river, or harbor from flooding. The prototype must be at least five inches high and wide enough to prevent the water on one side of a plastic container from flooding into the other side of the container. Each group receives a plastic container in which to build the prototype levee. Each team receives $10 (fun money) to purchase levee supplies from the list of approved materials. In small groups, answer the following: What is the problem to solve? What needs to be designed and who is it for? What are the project requirements and limitations?

Step 2: Research and design — Practice of planning and carrying out investigations

Students collaborate to brainstorm ideas and develop as many solutions as possible. Research to find out how real levees are constructed. Select at least two of the approved materials from the list and test how well they slow down water. Put a small hole in a paper cup, and put your test material at the bottom of the cup. Measure a specific amount of water in another cup. Then pour the water into the cup with a hole and observe how well the material absorbs the water. Record your observations. When engineers brainstorm ideas, they are open to many creative ideas—the more creative the idea, the better! Discuss with your group some ideas for how to build your levee. Write down or sketch every idea suggested and discussed. Small groups of students are presented with the problem and the first set of data (see Table 8.7).

Table 8.7. Data set 1 for StEMT activity for Lesson LS2.C

Company	Initial Cost	Yearly Maintenance Cost	Building Material	Time to Construct	Disruption to Local Community
Company A	$3 million	$2 million	Wood	6 months	Moderate
Company B	$4.5 million	$1 million	Plastic	11 months	Low
Company C	$7.8 million	$0.5 million	Stone	12 months	High
Company D	$6.2 million	$0.5 million	Concrete	18 months	High

Students will determine, based on the data, which materials should be used to build the levee using the following questions as a guide:

- What does the local government want from you?

- What are the factors that the local government want you to consider?

- What type of levee is least expensive to build? Why is this not necessarily the best option?

- What type of levee causes the least amount of disruption to the local community? Why is this not necessarily the best option?

Step 3: Plan—Practice of constructing explanations and designing solutions

Students will compare the best ideas, select one solution and make a plan to investigate it. Read through your ideas again, and, as a group, choose the concept that you think will work best.

Table 8.8. Budget chart for StEMT activity for Lesson LS2.C

Approved Materials	Cost	Amount Requested	Cost
One cup of sand or gravel	$1		
Five cotton balls	$1		
Straws	$1		
Craft sticks	$1		
One foot of duct tape	$2		
One sheet of paperboard	$2		
One plastic bag	$2		
One sponge	$2		
		TOTAL	

Describe or sketch the idea. Use Table 8.8 to determine how you will spend your budget on materials.

Students develop a procedure for choosing the best type of levee and rank their choices (1–4). Students use their procedure to choose the best levee and provide justifications and explanations for each ranking. Consider the following questions:

- How will you determine which material is best for the levee when you consider factors such as cost, cost of upkeep, the time it will take to build it, and the amount of disruption it will cause the local community?

- If cost were not a factor, which material would be best? Why?

- If impact on the community were not a factor (i.e., all types of levees caused minimal disruption), which material would you build the levee out of? Why?

- How will you determine which material is best for the levee, when you consider factors such as cost, cost of upkeep, the time it will take to build it, and the amount of disruption it will cause the local community?

Table 8.9. Data set 2 for StEMT activity for Lesson LS2.C

Company	Initial Cost	Yearly Maintenance Cost	Building Material	Time to Construct	Disruption to Local Community	Total Game Fish Population Downstream*	Flow**
Company A	$3 Million	$2 Million	Wood	6 months	Moderate	100,000	Very high
Company B	$4.5 Million	$1 Million	Plastic	11 months	Low	36,000	Low
Company C	$7.8 Million	$0.5 Million	Stone	12 months	High	85,000	High
Company D	$6.2 Million	$0.5 Million	Concrete	18 months	High	13,000	Very Low

*This category refers to what the total game fish population would be **AFTER** construction of this type of levee. Certain levee types would leave more game fish in the river; others would cause the game fish population to decrease significantly.

**This category refers to how much water will be able to flow through the levee. A high flow means water will still flow past the levee and will cause minimal disruption to the ecosystems on the other side of the levee. A low flow will cause very little water to flow through the levee but will cause much greater disruption (because they're losing a water source). However, a high flow indicates a less effective levee, while a low flow indicates a more effective levee.

Updated Problem

The government has received new information from a state environmental group that the construction of the levee could have an effect on the local game fish population. Fishing is an important part of the local economy and must be considered when building the levee. Analyze the new set of data that includes the impact the construction of a levee will have on the game fish population. Different levee types will also produce a different speed of water flow, which is another factor to consider. See Table 8.9.

As a group, students brainstorm how to determine the best type of levee to build based on the new information. They will use the guiding questions to assist them.

- Students will develop a procedure for choosing the best type of levee with the new data.

- Students will use their procedure to choose the best levee and provide justifications and explanations for each. They will rank their choices.

- Students will write their final recommendation letter to the client. They will detail and justify all of their recommendations.

After receiving Data Set 2, students will interpret the new data and determine if the materials they originally selected should be used to build the levee using the following questions as a guide:

- How do the new data change the way you think about the problem?

- How will you revisit your original procedure?
- Did you develop a new strategy to determine the best type of levee to build?
- Did you recommend a new material type when considering the new data?
- If you changed your recommendation, what influenced your decision? Did you use more than one strategy or procedure to determine the best material with which to build the levee?

Step 4: Create—Practice of developing and using models

Purchase materials and build your levee prototype. With the teacher's help, test the levee by pouring water into one side of the container. Describe what happened.

EVALUATE

- How will students demonstrate that they have achieved the lesson objective?
- Evaluation should be embedded throughout the lesson as well as at the end of the lesson.

Formative Assessment

Administer two formative assessments: the first formative assessment after students have been presented with the original problem and Data Set 1 and the second formative assessment after the groups have received the new data set (see Table 8.10).

Step 5: Test and Improve—Practice of obtaining, evaluating, and communicating information

Does the prototype work and does it solve the need? Students communicate the results and get feedback. Students then talk about what works, what does not, and what could be improved.

After engineers test their prototypes, they think about how well the prototypes worked. This helps them make changes to improve the final, real version. What did you like best about your levee system design? What changes would you make to your levee system if you were to build it again?

Students should present in small groups the results of their levee project and the changes they made.

Figure 8.10. Sample writing rubric for summative assessment for Lesson LS2.C

Content	3	2	1
Content accuracy	The letter accurately describes three limiting factors of the ecosystem. The letter accurately describes each of their effects on the local game fish population.	The letter accurately describes three limiting factors of the ecosystem. However, the letter incorrectly describes at least one of their effects on the local game fish population.	The letter accurately describes one limiting factor of the ecosystem. Other limiting factors are either incorrectly addressed or not addressed at all. The letter incorrectly describes their effects on the local game fish population.
Evidence to support claim	Evidence for the recommendation is clear. There are three detailed explanations supporting the claim.	Evidence for the recommendation is clear. There are one or two detailed explanations supporting the claim.	A recommendation to support the claim is given, but there is no evidence to support.
Grammar, spelling, and organization	Writer makes no errors in grammar or spelling. The letter has a logical organization and flow.	Writer makes one or two errors in grammar or spelling. The letter is logical, but there are unclear transitions between ideas.	Writer makes three or more errors in grammar or spelling. The letter demonstrates a lack of organization and has no transitions between ideas.

Designing the Super Explore

Life Sciences (LS3): Heredity: Inheritance of Traits (LS3.A.) and Variation of Traits (LS3.B.)

Teacher Brief

Imagine the Teenage Mutant Ninja Turtles meets X-Men. Designing the Super Explore is an engineering design challenge added to an inheritance lesson. Dependent on the grade level and purpose of this lesson, students may need some background knowledge of the terms *chromosome*, *allele*, *dominant*, and *recessive,* and possibly a review of cell division. The Explore section of the lesson has students (divided in pairs) choosing from prepared bags of beads. This random sampling provides students with an understanding of how different traits develop and are expressed (dominant) or not expressed (recessive). Within the Explain sections, terms such as *genotype, phenotype, homozygous,* and *heterozygous* are introduced and used in context. Please note that this lesson focuses only on simple on simple (dominant or recessive) Mendelian genetics. Students need to know that this is a simplified model of genetics.

The more traditional elaboration for a lesson of this type is to make it relevant by looking at traits exhibited by the students' parents. Students will determine the probability of those potential inherited traits. The StEMT*ified* version uses a problem that needs to be solved by genetic engineering to create a super-organism capable of completing a task. Like most real-world problems, genetic engineering may come with some sacrifice and, depending on the class, ethics may also be a topic that surfaces during the lesson. The students will have to make decisions about which genes will need to be expressed by the genetically engineered organism based on a defined problem with specific parameters and limitations. We have provided one example, but there are many other possible problems that would allow students to have further interactions with the concept of inheritance and variation of traits. We urge you to be creative when coming up with the task or have students use their imaginations to create the task to be solved.

Prior Knowledge

By the end of grade 5. Offspring acquire a mix of traits from their biological parents. Different organisms vary in how they look and function because they have different inherited information. In each kind of organism, there is variation in the traits themselves, and different kinds of organisms may have different versions of the trait. The environment also affects the traits that an organism develops—differences in where they grow or in the food they consume may cause organisms that are related to end up looking or behaving differently. (*Framework*; NRC 2012, pp.158, 160).

LS3.B: By the end of grade 8. In sexually reproducing organisms, each parent contributes half of the genes acquired (at random) by the offspring. Individuals have two of each chromosome and hence two alleles of each gene, one acquired from each parent. These versions may be identical or may differ from each other (*Framework;* NRC 2012, pp.158, 160).

LS3.A: By the end of grade 5. Many characteristics of organisms are inherited from their parents. Other characteristics result from individuals' interactions with the environment, which can range from diet to learning. Many characteristics involve both inheritance and environment (*Framework*; NRC 2012, p. 158).

LS3.A: By the end of grade 8. Genes are located in the chromosomes of cells, with each chromosome pair containing two variants of each of many distinct genes. Each distinct gene chiefly controls the production of a specific protein, which in turn affects the traits of the individual (e.g., human skin color results from the actions of proteins that control the production of the pigment melanin). Changes (mutations) to genes can result in changes to proteins, which can affect the structures and functions of the organism and thereby change traits. In sexually reproducing organisms, each parent contributes (at random) half of the genes acquired by the offspring. Individuals have two of each chromosome and hence two alleles of each gene, one acquired from each parent. These versions may be identical or may differ from each other (*Framework*; NRC 2012, p. 158).

Framework Questions: How are the characteristics of one generation related to the previous generation (LS3.A)? Why do individuals of the same species vary in how they look, function, and behave (LS3.B)?

StEMTified Question: Why is understanding specific genetic traits (genotype) and how they present (phenotype) important for bioengineering? How can public concerns be minimized when dealing with genetically engineered plants and animals?

Dimension 1: Practices (What should students be doing?)

- Developing and using models
- Analyzing and interpreting data
- Constructing explanations and designing solutions
- Engaging in argument from evidence

Dimension 2: Crosscutting Concepts

- *Structure and function.* The way in which an object or living thing is shaped and its substructure determine many of its properties and functions. Complex and microscopic structures and systems can be visualized, modeled, and used to describe how their function depends on the relationships among its parts.

Dimension 3: Disciplinary Core Idea for Life Sciences (LS3.B): Variation of Traits

Variation among individuals of the same species can be explained by both genetic and environmental factors. Individuals within a species have similar but not identical genes. In sexual reproduction, variations in traits between parent and offspring arise from the particular set of chromosomes (and their respective multiple genes) inherited, with each parent contributing half of each chromosome pair. More rarely, such variations result from mutations, which are changes in the information that genes carry. Although genes control the general traits of any given organism, other parts of the DNA and external environmental factors can modify an individual's specific development, appearance, behavior, and likelihood of producing offspring. The set of variations of genes present, together with the interactions of genes with their environment, determines the distribution of variation of traits in a population (*Framework*; NRC 2012, p. 160).

Instructional Duration: Four to five 50-minute classes

Materials per Class of 30

- 38 small brown paper lunch bags per group
- Pipe cleaners cut in half
- Plastic beads (the number of plastic beads depends on the number of students in the class)
    - About 80 red and 70 purple plastic beads (represent tongue rolling alleles)
    - About 100 black and 50 white plastic beads (represent eyelash length alleles)
    - About 90 brown and 60 green plastic beads (represent eye color alleles)
    - About 80 orange and 70 yellow plastic beads (represent ear lobe attachment alleles)
    - Pink and blue beads to represent gender

Teacher Preparation

Prepare two bags of each of the following mixtures. Label one bag *mother* and one *father*. These resemble the ratio occurring in nature.

- About 80 red/70 purple plastic beads—labeled "tongue rolling"
- About 100 black/50 white—labeled "eyelash length"
- About 90 brown/60 green —label "eye color"
- About 80 orange/70 yellow —label "earlobes"

Safety Notes

1. Handle sharps (pipe cleaners) with caution, as they can cut or puncture skin.
2. Immediately pick up any beads off the floor.
3. Wash hands with soap and water upon completing this activity.

Guiding Question: How can genotype and phenotype be determined?

Misconceptions

- Females inherit their traits from their mom, and males inherit their traits from their dad.

- A dominant trait is the one most likely to be found in the population.

ENGAGE

- How will you capture students' interest and reveal misconceptions?

- What kinds of questions should the students ask themselves after the engagement?

Refer to a picture of a family. Students work with shoulder partners (think–pair–share) to answer the following two questions. Randomly call on students and facilitate discussion. Do NOT correct misconceptions.

- What features do the children have that resemble their mother? Their father?

- Why don't children look exactly like their same-sex parent?

EXPLORE

- Describe the hands-on activities that will provide experience of the phenomenon.

- List "big idea" conceptual questions you will use to encourage and focus students' exploration and allow the students to test their ideas.

Prior Knowledge

Prompt for descriptions of the terms *chromosome*, *allele*, *dominant*, and *recessive*. Explain that students will be investigating the inheritance of the following four human traits (optional: display pictures that show each trait).

- **Tongue rolling:** Red = dominant, rolling (allele = R); Purple = recessive, nonrolling (allele = r)

- **Lashes:** Black = dominant, long (allele = L); White = recessive, short (allele = l)

- **Eyes:** Brown = dominant, brown (allele = B); Green = recessive, green (allele = b)

- **Earlobes:** Orange = dominant, detached (allele = D); Yellow = recessive, attached (allele = d)

- **Gender:** Father (male)—one pink, one blue (XY); Mother (female) = two pink (XX)

Review meiosis and how gametes end up with one half the number of chromosomes. Ask: How many alleles do you have for each trait? Where do they come from? Explain that the chart shows which trait each color represents. Explain that the "case" of the letter is important as it represents dominant (capital) or recessive (lowercase). This will be discussed later.

Divide students into pairs and explain that one will represent the "mother" and one will represent the "father". Have students draw "father" and "mother" cards out of a container to determine "mother" and "father." Display the list from Table 8.11 showing the trait each color represents.

- "Mothers" choose two beads from each of the four prepared bags. The chosen beads from each bag represent a pair of alleles for each of the four traits. "Mothers" additionally take two pink beads, representing a female XX chromosome.

- "Fathers" do the same, but additionally choose a pink bead and a blue bead, representing the XY male chromosome.

- Parents, list your two-letter allele combinations for each of the four traits under "Genotype."

- One at a time for each trait, "mothers" place your two beads in a lunch bag and without looking, choose one to "give" to the baby. "Fathers" do the same.

- The chosen beads are combined and made into pipe cleaner chromosomes for the baby.

- Students record the two-letter allele combinations for each trait of the offspring (baby) under "Genotype" in their science notebook (see Table 8.11).

Table 8.11. Data collection chart for Explore activity for Lesson LS3.A and Lesson LS3.B

Trait	Father		Mother		Offspring	
	Genotype	Phenotype	Genotype	Phenotype	Genotype	Phenotype
Tongue rolling						
Lashes						
Eyes						
Earlobes						
Gender	XY	Male	XX	Female		

EXPLAIN

- Student explanations should precede your explanations or introduction of terms. What questions or techniques will you use to help students connect their exploration to the concept under examination?

- List higher-order thinking questions that you will use to solicit *student* explanations and to help students construct and justify their explanations.

Learners explain through analysis of their exploration, so their understanding is clarified and modified by the teacher using scientific terminology. Ask the following probing questions for small group discussion. Students record their answers in their student notebooks. (Discuss the meaning of prefixes: homo-, hetero-, pheno-, geno- to assist students with important vocabulary.)

- What do the letter combinations in your notebook represent, and what is the term used for these combinations? What is the term for the appearance of these two alleles?

- What is the term for two alleles that are the same (AA or aa)?

- What is the term for two alleles that are different (Aa)?

- Did your offspring have the exact same genotype (combination of alleles) for any of the traits as either of their parents? Why or why not?

- If a dominant allele is present (e.g. tongue rolling) what phenotype will be presented? Write the phenotype for each of the traits for parents and offspring in the chart.

- Did your offspring have the exact same phenotypes (appearance) as either of their parents? Why or why not? Based on the above, what do you think *dominant* means?

- A Punnett square is a chart used to determine the probability of offspring inheriting certain traits (see Table 8.12). Parent A, homozygous for brown eyes, is crossed with Parent B, with green eyes—what are the genotypes for these two parents for eye color? Demonstrate how the Punnett square is used to determine genotype and phenotype for specific traits.

  o The genotypes of one parent are written across the top of the chart.

  o The genotypes of the other parent are written down the side of the chart.

  o Each square inside the chart is filled in with one allele from each parent and represents the probability of the offspring receiving a specific genotype.

Table 8.12. Punnett square sample for Lessons LS3.A and LS3.B.

B		*Parent A Genotype <u>BB</u>*	
		B	
Parent B Genotype <u>bb</u>	*b*	**Bb**	**Bb**
	b	**Bb**	**Bb**

- ○ What percentage of the offspring will have brown eyes?
- ○ What percentage of the offspring will have green eyes?

ELABORATE

- Describe how students will develop a more sophisticated understanding of the concept.
- What vocabulary will be introduced, and how will it connect to students' observations?
- How is this knowledge applied in our daily lives?

Learners elaborate and solidify their understanding of the concept and apply it in a real-world situation, resulting in a deeper understanding. Students show work in their science notebooks.

- Create a Punnett square for each of the traits for your two parents.
  - ○ For each trait, list the three genotypes and the probabilities that the offspring will inherit those genotypes.
  - ○ For each trait, list the two phenotypes and the probabilities that the offspring will inherit those phenotypes.

Look-Fors
Students should engage in the following behaviors:

- *Practice of developing and using models.* Students will develop and use models (Punnett squares, diagrams, and simulations) to describe the cause and effect relationship of gene transmission from parents to offspring and the resulting genetic variation (SEP 2; MP 4).

- *Practice of obtaining, evaluating, and communicating information.* Students will apply scientific vocabulary after exploring a scientific concept (SEP 6; MP 7).

StEMT*ify*

Science: What real-world issues and problems exist in the area of genetic engineering?

Engineering: Is the STEM lesson guided by the engineering design process?

- Students should be immersed in hands-on inquiry and open-ended exploration.
- Students should be involved in productive teamwork.

Mathematics: Does the STEM lesson rigorously apply mathematics and science content that the students are learning?

Technology: Does the STEM lesson allow for multiple right answers through prototype development, testing, and improvement of design? Encourage students to monitor and evaluate their progress, change course if necessary, and ask, "Does this make sense?"

StEMT*ified* Question: Why is understanding specific genetic traits (genotype) and how they present (phenotype) important for bioengineering? How can public concerns be minimized when dealing with genetically engineered animals?

Bioengineering a Super-Organism
Step 1: Ask—Practice of asking questions (science) and defining problems (engineering)

Scientists have discovered harmful, mesopelagic, anaerobic bacteria that have begun to migrate to the abyssopelagic zone, where they can reproduce rapidly and uncontrollably because there is very little oxygen in this deep-water ocean zone. These bacteria emit a sweet-smelling odor when exposed to light. Students have been asked to help geneticists engineer a new species of organism that can survive on the bottom of an ocean abyss and will digest large quantities of the bacteria before they reproduces uncontrollably, however, the organism cannot have the ability to reproduce because it is genetically engineered. Students must select two parent organisms (Organism A, B, or C) that will produce a combination of alleles that would result in an organism that would meet the requirements to successfully decrease the bacteria from the deep ocean zone. What is the problem to solve? What needs to be designed? What is the goal? Review pertinent vocabulary (*dominant* versus *recessive*; *homozygous* versus *heterozygous*; *genotype* versus *phenotype*).

Step 2: Research and design—Practice of planning and carrying out investigations

For a basic overview of genetics, teachers and students may use the OER CK–12 Standards-Aligned FlexBook textbook, *Mendel's Laws and Genetics* at *www.ck12.org/biology/Mendels-Laws-and-Genetics*.

Small-group discussion

What are the project requirements and limitations? Which traits does each collaborative group feel is most important for the genetically engineered organism to be successful in decreasing the harmful bacteria population? Each group must justify their reasons for trait selection for the following genotypes:

- Bioluminescence (B) and no bioluminescence (b)
- Enhanced sight (E) and blindness (e)
- Enhanced hearing (H) and deafness (h)
- Large mouth (M) and small mouth (m)
- Enhanced smell (S) and no smell (s)
- Fertile (Y) and sterile (y)

Step 3: Plan—Practice of constructing explanations and designing solutions

Parent Organism A

Homozygous dominant for bioluminescence
Homozygous dominant for enhanced sight
Homozygous dominant for enhanced hearing
Homozygous dominant for large mouth
Homozygous dominant for enhanced smell
Homozygous dominant for fertility

Parent Organism B

Heterozygous for bioluminescence
Heterozygous for enhanced sight
Heterozygous for enhanced hearing
Heterozygous for large mouth
Heterozygous for enhanced smell
Heterozygous for fertility

Parent Organism C

Homozygous recessive for bioluminescence
Homozygous recessive for enhanced sight
Homozygous recessive for enhanced hearing
Homozygous recessive for large mouth
Homozygous recessive for enhanced smell
Homozygous recessive for fertility

- What things do you need to include in your solution? [*Reasons for selected genotypes for each parent fish (A, B, or C) and the process the team used to determine the selections.*]

Step 4: Create—Practice of developing and using models

Students use Punnett squares to show the genotypes for the two selected parent organisms with possible allele combinations for the new organism, including the phenotypes for each trait.

- Is there an alternative solution to what the client is asking? Why or why not?

Step 5: Test and improve—Practice of obtaining, evaluating, and communicating information

- What if there were two of each parent organism? Would the selection of traits change?

Test, evaluate, and revise the genetically engineered organism as necessary using the new information (e.g., two parent organisms with A genotype/phenotype) and provide the reasoning for any modifications. Teams should begin preparing their presentations.

Look-Fors

Students should engage in the following practices:

- *Practice of asking questions (science) and defining problems (engineering).* Students will ask questions, define problems, and predict solutions/results (SEP 1; MP 1).

- *Practice of planning and carrying out investigations.* Students will be actively engaged and work cooperatively in small groups to complete investigations, test solutions to problems, and draw conclusions. Use rational and logical thought processes, and effective communication skills (writing, speaking and listening; SEP 7, SEP 8; MP 3).

- *Practice of constructing explanations and designing solutions.* Students will design, plan and carry out investigations to collect and organize data (SEP 3; MP 1).

- *Practice of obtaining, evaluating, and communicating information.* Students will analyze and interpret data to draw conclusions and apply understandings to new situations (SEP 4; MP 5).

EVALUATE

- How will students demonstrate that they have achieved the lesson objective?

- Evaluation should be embedded throughout the lesson as well as at the end of the lesson.

Students should give presentations on gene combination, including the Punnett squares to explain the decision making process for genetically engineered fish. This could be in the form of RAFT writing (RAFT = **R**ole, **A**udience, **F**ormat, **T**ask). An example of a RAFT is the **Role** is bioengineer, the **Audience** is geneticists requesting assistance, the **Format** is a scientific paper, and the **Task** is to make a claim for two parent organisms resulting in the most well-adapted offspring for the engineering design. The evidence to support the claim could be the Punnett squares showing genotype and phenotype of offspring with targeted traits and reasoning for the selected traits (see the CER Rubric in Appendix).

Summative Assessment: Write about it!

Why don't children look identical to their same-sex parents? Support your claim with evidence from the Explore, Explain, and Elaborate sections. Provide scientific data and use of content-specific vocabulary to support your claim.

It's a What?

Life Sciences: Biological Evolution: Unity and Diversity— Natural Selection (LS4.B)

Teacher Brief

Students are fascinated by strange-looking animals that have adapted to live in environments different from our own. The mainstay of most of these lessons is either research or a sorting activity, which can be rather passive in nature. In our StEMT*ify* extension, students are actively engaged by solving a problem that is realistic: designing an isolated area where a specific animal, with specific needs, can live for an extended length of time. We have included a list of questions that students should consider as they design their enclosures, but teachers may have even more questions. Our goal is to take a passive lesson and provide a challenge for the students that will use what they have learned, perhaps force them to do more research, and require them to consider factors that are important to certain species with particular adaptations.

We have seen this done in classrooms where students design their enclosures by developing a two-dimensional model on paper, while in other classrooms teachers provided students with materials to create a three-dimensional model. In all cases, it is important for the teacher to continuously ask questions to force students to consider every aspect of their enclosure and how it meets the needs of the animal. We feel that sharing out with peers during the design process is also critical. Students are so proud of their designs that they want to share their ideas, and doing so helps solidify the knowledge construction we hope the lesson facilitates.

Prior Knowledge

By the end of grade 5. Sometimes the differences in characteristics between individuals of the same species provide advantages in surviving, finding mates, and reproducing (*Framework;* NRC 2012, p. 164).

By the end of grade 8. Genetic variations among individuals in a population give some individuals an advantage in surviving and reproducing in their environment. This is known as natural selection. It leads to the predominance of certain traits in a population and the suppression of others. In *artificial selection*, humans have the capacity to influence certain characteristics of organisms by selective breeding. One can choose desired parental traits determined by genes, which are then passed on to offspring (*Framework;* NRC 2012, p. 164).

Framework **Question:** How does genetic variation among organisms affect survival and reproduction?

StEMT*ified* Question: Why is it important for scientists to understand adaptations specific to a population and how those adaptations are essential for survival in specific habitats (enclosures)?

Dimension 1: Practices (What should students be doing?)

- Constructing explanations and designing solutions
- Engaging in argument from evidence
- Obtaining, evaluating, and communicating information

Dimension 2: Crosscutting Concepts

- *Cause and effect: Mechanism and explanation.* Events have causes—sometimes simple, sometimes multifaceted. A major activity of science is investigating and explaining causal relationships and the mechanisms by which they are mediated. Such mechanisms can then be tested across given contexts and used to predict and explain events in new contexts.

- *Structure and function.* The way in which an object or living thing is shaped and its substructure determine many of its properties and functions.

Dimension 3: Disciplinary Core Idea

Genetic variation in a species results in individuals with a range of traits. In any particular environment, individuals with particular traits may be more likely than others to survive and produce offspring. This process is called natural selection and may lead to the predominance of certain inherited traits in a population and the suppression of others. Natural selection occurs only if there is variation in the genetic information within a population that is expressed in traits that lead to differences in survival and reproductive ability among individuals under specific environmental conditions. If the trait differences do not affect reproductive success, then natural selection will not favor one trait over others (*Framework;* NRC 2012).

Instructional Duration: Approximately five 45-minute class periods, depending on student level and background knowledge. Timing for each section will vary.

Materials per Class of 30

- Pictures of Amazonian royal flycatcher, babirusa, fossa, pink fairy armadillo, gerenuk, lamprey, Irrawaddy dolphin, and maned wolf (projected or in sets per group)

- Pictures of animals living in habitats (desert, temperate forest, tundra, etc.)

- For a basic overview of natural selection, teachers may use the OER CK–12 Standards-Aligned FlexBook textbook *Life Science (www.ck12.org/teacher).*

- Article from *Popular Science* on bedbug evolution: Borel, B. (2011). New tougher bedbugs are harder than ever to kill. Popular Science. Retrieved from *www.popsci. com/science/article/2011-05/bedbugs-are-harder-ever-kill*

Guiding Questions

Can different plants and animals have different adaptations? Why or why not? What are some adaptations of plants or animals and how do these adaptations aid in survival in a specific environment? What would happen if an animal is not able to adapt to its environment?

Misconceptions (AAAS Project 2061, n.d.)

- Individual organisms can deliberately develop new heritable traits because they need them for survival.

- Change occurs in the inherited characteristics of a population of organisms over time because of the use or disuse of a particular characteristic.

- The internal chemistry, appearance, and behavior of a species do not change, even over long periods of time.

- Changes to the environment cannot lead to changes in the traits of species living in that environment.

ENGAGE

- How will you capture students' interest and reveal misconceptions?
- What kinds of questions should the students ask themselves after the engagement?

After a half-century of relative inactivity in the United States, bedbugs returned in the late 1990s. In the 1950s, use of DDT and other insecticides almost eliminated the pests. Scientists hypothesize that the few bedbugs that survived reproduced and passed along pesticide-resistant adaptations to their offspring, including thicker, denser, wax-like exoskeletons that repel chemicals in pesticides and a faster metabolism that produces chemical blockers rendering pesticides harmless (Borel 2011).

- What is an adaptation? How do adaptations aid in an organism's survival?

Show students the pictures of animals and plants that live locally. Elicit student ideas for what physical characteristics or adaptations help these animals and plants survive in their environment. Ask the following questions to elicit discussion:

- How has the plant or animal adapted to survive in the environment?"
- Why might [insert characteristic] be important for living in a tropical rainforest? Would this same characteristic help in a desert? Why or why not?

Have students look at some of the most unusual animals (e.g., Amazonian royal flycatcher, babirusa, fossa, pink fairy armadillo, gerenuk, lamprey, Irrawaddy dolphin, maned wolf) and discuss in small groups what type of environment they think the animals would be best adapted to and why.

- Which animal do you think is the most unusual and why? What are some physical characteristics of this animal and how do these adaptations help it survive?
- What type of environment might this animal be best adapted for? How do you know? Why do physical characteristics (adaptations) of animals in different habitats vary from one animal to another?

EXPLORE

- Describe the hands-on activities that will provide experience of the phenomenon.
- List "big idea" conceptual questions you will use to encourage and focus students' exploration and allow the students to test their ideas.

Prior Knowledge

Students should recognize that there are different climate zones (polar, temperate, and tropical) and different habitats (desert, tundra, grassland, temperate forest, etc.) in the world and that habitats and climate have specific characteristics.

Have students choose an animal or plant and discuss the following questions in small groups:

- In what climate does the animal live? What adaptations does the animal possess that allow it to survive in the environment?

- What are some ways that an animal might struggle for survival in its environment?

- What are some ways that a plant might struggle for survival in its environment?

- What are some ways adaptations help organisms survive and reproduce?

- What adaptations would help an animal or plant living in the different habitats to survive (show pictures of animals that live in the desert, tundra, grassland, temperate forest, etc.)?

EXPLAIN

- Student explanations should precede your explanations or introduction of terms. What questions or techniques will you use to help students connect their exploration to the concept under examination?

- List higher-order thinking questions that you will use to solicit *student* explanations and to help students construct and justify their explanations.

Have students conduct research on a plant and animal of their choosing. For each, the student will need to identify four adaptations and examples of how the adaptations help the plant or animal survive in the specific habitat or climate zone. Design a poster of the animal or plant's adaptations for a class gallery walk.

ELABORATE

- Describe how students will develop a more sophisticated understanding of the concept.

- What vocabulary will be introduced, and how will it connect to students' observations?

- How is this knowledge applied in our daily lives?

Students will participate in a gallery walk to learn about the research their classmates conducted on their selected plant or animal. Have students take notes in their science journals during the gallery walk, focusing on the following guiding questions: Can different plants and animals have different adaptations? Why or why not? What are some adaptations of plants or animals and how do these adaptations aid in survival in a specific environment? What would happen if an animal was not able to adapt to its environment?

Compare and contrast the adaptations encountered during the gallery walk (e.g., pick two organisms and complete a Venn diagram, or pick three organisms and complete a triple Venn).

StEMT*ify*

Science: What real-world issues and problems exist in the area of adaptation and survival of species?

Engineering: Is the STEM lesson guided by the engineering design process?

- Students should be immersed in hands-on inquiry and open-ended exploration.

- Students should be involved in productive teamwork.

Mathematics: Does the STEM lesson rigorously apply mathematics and science content that the students are learning?

Technology: Does the STEM lesson allow for multiple right answers through prototype development, testing, and improvement of design? Encourage students to monitor and evaluate their progress, change course if necessary, and ask, "Does this make sense?"

StEMT*ified* Question: Why is it important for scientists to understand adaptations specific to a population and how these adaptations are essential for survival in specific enclosures?

A conservatory has been given the task of captive breeding of animal species recently discovered that were thought to be extinct. These animals must be placed in quarantine for six months before being released into the local zoo habitats for the captive breeding program. Using what you know about animal adaptations necessary for survival, design a zoo-like habitat (enclosure) for the animals that will be best suited to their adaptations and survival. Asks students to discuss in small groups: What is quarantine, and why is it important when dealing with animals from different countries?

Step 1: Ask—Practice of asking questions (science) and defining problems (engineering)

What is the problem to solve? What needs to be designed, and who is it for? What is the goal?

- What information has been given to solve this problem? What environments are available?

- What size should an animal be (in kilograms) to be placed in an indoor enclosure or aquarium? What other characteristics might influence where the animal should be placed?

- What are the most important things to consider in your procedure? Can this procedure be used with a new set of animals?

Step 2: Research and design—Practice of planning and carrying out investigations

What are the project requirements and limitations? (Use Tables 8.13 [p. 133] and 8.14 [p. 134.])

Chart with characteristics of animals. Based on the physical characteristics (adaptations) about the animals from the conservatory, determine the type of habitat each animal will need based on its adaptations. Be able to explain to other groups why you placed each animal in its habitat so the method can be repeated if more animals arrive at the local zoo from other countries. Revise rules for animal placement based on small-group discussions.

Table 8.13. Chart for StEMT activity for Lesson LS4.B

Animal	Body Covering	Color	Feet	Mouth	How It Moves	Protection	Diet	Size
1	Thick, heavy fur	White	Furry, large, flat	Sharp pointy teeth	Walks, runs	Sharp teeth and claws	Carnivore	1,000 kg
2	Smooth fur	Light brown	Small, hand-like	Tiny teeth	Climbs, walks	Hides, climbs small branches	Herbivore	23 g
3	Smooth, moist skin	Green/brown	Webbed	Large, ridges on jaw, no teeth	Swims	Slippery, poisonous if eaten	Carnivore, feeds in water	362 kg
4	Thick, hard scales	Blue/white	Scales on bottom of wide, flat feet	Small jaw, sharp teeth	Walks, runs	Armor plating	Omnivore	2,000 g
5	Exo-skeleton	Red/purple	Claws	Beak	Flies	Sharp beak and claws, flies away	Carnivore	9,000 g
6	Exo-skeleton	Orange/brown	Hairy	Small	Crawls and hops	Stinger	Herbivore	1 kg
7	Scales	Green	Hoof-like	Flat teeth	Walks	Runs away	Herbivore	114 kg
8	Scales	Silver/blue	No feet	Small, sharp teeth	Swims	None	Omnivore	2.5 kg

Table 8.14. Data collection chart for StEMT activity for Lesson LS4.B

Write the number of the animal to identify the type of habitat/enclosure in which it belongs.				
Types of Enclosures/ Habitats	**Large Fenced Area**	**Enclosed Outdoor Area With Chain-Link Fence Roof**	**Cage Inside Building**	**Aquarium**
Desert				
Arctic				
Rainforest				
Temperate forest				
Grassland				
Swamp				
Freshwater				

Step 3: Plan—Practice of constructing explanations and designing solutions

Guiding questions as students work in small groups:

- What characteristics would make an animal well adapted to living in each habitat? Could an animal with thick, heavy fur live in the desert? Why or why not? Could an animal with smooth, moist skin survive in the Arctic? Why or why not?

- What characteristics would determine if an animal needs to live in a large or small area?

- Why would an animal need to be kept in an indoor or outdoor enclosure? An aquarium?

- Why did you choose to place an animal in a specific habitat? Could the animal live elsewhere? Describe how the habitat is appropriate for the animal's adaptations.

- Would all animals with that characteristic be suited for that environment? Why or why not?

Step 4: Create—Practice of developing and using models

The EDGE of Existence program highlights and conserves one-of-a-kind species that are on the verge of extinction. Evolutionarily distinct and globally endangered (EDGE) species have few close relatives and are often extremely distinct in the way they look, live and behave. These unique species are also on the verge of extinction (www.edgeofexistence.org).

In pairs, students will select one species identified on the EDGE website, conduct research on the animal's adaptations, and design an enclosure based on the animal's unique adaptations. Explain how the enclosure is suited for the animal's adaptations and what habitat/climate the enclosure mimics (simulates). Create a public service announcement (PSA) to post on the enclosure to bring awareness to conservation of the species.

Step 5: Test and Improve—Practice of obtaining, evaluating, and communicating information

- Could there be more than one habitat for each animal? Why or why not?

- Will this procedure work for all animals with similar characteristics? Why or why not?

EVALUATE

- How will students demonstrate that they have achieved the lesson objective?

- Evaluation should be embedded throughout the lesson and at the end of the lesson.

Formative Assessment

Students should be able to justify why each animal belongs in a certain habitat based on the animal's adaptations. Present the species PSA to the class. Have students brainstorm what should be included in the PSA so everyone is consistent with the PSA message and aware of the rubric on which the PSA will be graded. This also gives students ownership of the task.

Summative Assessment: Write about it!

Explain why adaptations displayed by the animals from Australia and Siberia enable them to survive in different environments. Describe your favorite animal. Describe why this animal is adapted to its habitat. Using the CER rubric in Appendix A, students should use evidence from the Explore and StEMT*ify* sections to justify their reasoning.

References

American Association for the Advancement of Science (AAAS). 1993. *Benchmarks for science literacy.* New York: Oxford University Press.

Borel, B. 2011. New tougher bedbugs are harder than ever to kill. *Popular Science.* Retrieved from *ww.popsci.com/science/article/2011-05/bedbugs-are-harder-ever-kill*

Keeley, P. 2007. *Uncovering student ideas in science, Volume 2.* Arlington, VA: NSTA Press.

National Research Council (NRC). 2012. *A framework for K–12 science education: Practices, crosscutting concepts, and core ideas.* Washington, DC: National Academies Press.

CHAPTER
9 | Physical Sciences StEMT Lessons

Is It Air or Wind?

Physical Sciences: Structure and Properties of Matter (PS1.A)

Teacher Brief

As humans, we live in a sea of air to the point that we forget the air is there. The fact that air is matter is a difficult concept for the young mind. The lesson we chose brings in some of our favorite hands-on activities. It starts with the classic three-holed bottle activity, which engages students and gets them to consider air as having the ability to cause change. These demonstrations (dry ice and water bottle) are must-dos … the students will be amazed! The Explore section has students making their own thermometer—demonstrating that air is matter and expands and contracts with increases in energy (i.e., temperature). During this portion, we typically use a coffee maker to heat the water, eliminating the need for young students to have a heat source. For older students, this same activity can be used with a variety of temperatures, as students will have the ability to warm the water themselves with a heat source provided for each group. In either case, the students will notice the increase in the size of the balloon, which demonstrates the ability of air to provide a force. One lesson we have learned in doing this activity is to not make the balloon too big. Also, make sure that there is time provided for the air inside of the test tube to affect the attached balloon.

During the Explain section there are ample opportunities to bring in graphing skills. Older students can even do interpolations and extrapolations based on their collected data. There are also many opportunities to include discussions of other important nature of science concepts.

The engineering design challenge is meant to continue the process of learning that air has the ability to create a change (i.e., provide a force). This activity can be as easy or complex as you want to make it. While making the windmill, one option to incorporate more mathematics is for students to purchase materials to build the windmill. This option brings in some relevance to real-world scenarios, as science knowledge is always constructed within societal norms.

As far as the challenge of hauling up an object of mass with a string wrapped around the windmill's shaft, the mass of the object will be depend on the materials used. We typically choose a very light mass, so students can be successful. If time is not a constraint, a fun extension is to have the mass that is being lifted to be part of the challenge. Consideration of the size (circumference) of windmill will need to be made for this to be a fair test. Again, there are many opportunities to engage students in the practice of science concepts throughout this task. A special note about this engineering design challenge is that it can easily be changed to water for the supplying force, depending on standards.

By the end of grade 5. Matter of any type can be subdivided into particles that are too small to see, but even then the matter still exists and can be detected by other means (e.g., by weighing or by its effects on other objects). For example, a model showing that gases are made from matter particles that are too small to see and are moving freely around in space can explain many observations, including the inflation and shape of a balloon; the effects of air on larger particles or objects (e.g., leaves in wind, dust suspended in air); and the appearance of visible-scale water droplets in condensation, fog, and, by extension, also in clouds or the contrails of a jet. The amount (weight) of matter is conserved when it changes form, even in transitions in which it seems to vanish (e.g., sugar in solution, evaporation in a closed container). Measurements of a variety of properties (e.g., hardness, reflectivity) can be used to identify particular materials (*Framework*; NRC 2012).

Framework Question: How do particles combine to form the variety of matter one observes?

StEMT*ified* Question: How do windmills support the claim that matter is everywhere and can be too small to be seen?

Dimension 1: Practices (What should students be doing?)

- Developing and using models
- Using mathematics, information and computer technology, and computational thinking
- Constructing explanations and designing solutions
- Engaging in argument from evidence

Dimension 2: Crosscutting Concepts

- *Scale, proportion, and quantity.* In considering phenomena, it is critical to recognize what is relevant at different measures of size, time, and energy and to recognize how changes in scale, proportion, or quantity affect a system's structure or performance. Natural objects exist from the very small to the immensely large. Standard units are used to measure and describe physical quantities such as weight, time, temperature, and volume. Students develop a model to describe the concept that matter is made of particles too small to be seen.

- *Systems and system models.* Defining the system under study—specifying its boundaries and making explicit a model of that system—provides tools for understanding and testing ideas that are applicable throughout science and engineering. Science assumes consistent patterns in natural systems.

Dimension 3: Disciplinary Core Idea

The existence of atoms, now supported by evidence from modern instruments, was first postulated as a model that could explain both qualitative and quantitative observations about matter (e.g., Brownian motion, ratios of reactants and products in chemical reactions). Matter can be understood in terms of the types of atoms present and the interactions both between and within them. The states (i.e., solid, liquid, gas, or plasma), properties (e.g., hardness, conductivity), and reactions (both physical and chemical) of matter can be described and predicted based on the types, interactions, and motions of the atoms within it. (*Framework*; NRC 2012).

Misconceptions

- Students think that larger objects always have more mass.
- Students think that mass and weight are the same.

Instructional Duration: Three to four 45-minute class periods

Materials per Class of 30

Small chunk of dry ice for demonstration, nine-inch balloon, ice, graduated cylinder, 1,000 ml beaker, water balloon, thermometer, ruler, dry-erase marker, coffee pot or microwave to heat water (water should only be heated to the temperature of warm coffee to minimize the risk of burns)

Engineering Design Challenge

For class: Hair dryer or fan; small object for each team to lift (e.g. toy car, yogurt cup filled with a few coins, tea bag, battery, pencil)

Per group: Wooden stick, wooden spoons, small wood (balsa) pieces, bendable wire, string, paper clips, rubber bands, toothpicks, aluminum foil, tape, dowels, glue, paper, cardboard, plastic wrap, indirectly vented chemical splash goggles, nonlatex gloves, nonlatex aprons

Safety Notes

1. Personal protective equipment (splash goggles, gloves, and aprons) should be worn during the setup, hands-on, and takedown segments of the activity and demonstration.

2. Use caution when working with sharps (toothpicks, clips, wire, etc.) that can cut or puncture skin.

3. Use caution when heating and handling hot substances.

4. Keep electrical devices (coffee pot, microwave, etc.) away from water sources (shock hazard).

5. Use caution when handling glassware (can shatter and puncture skin).

6. Never handle dry ice without proper thermal gloves. Always use a safety shield for a dry ice demonstration.

7. Immediately wipe up any water on the floor (slip-and-fall hazard).

8. Wash hands with soap and water upon completing this activity.

For a basic overview of physical science, teachers may use the OER CK–12 Standards-Aligned FlexBook Textbook *Physical Science* (*www.ck12.org/teacher*).

ENGAGE

- How will you capture students' interest and reveal misconceptions?
- What kinds of questions should the students ask themselves after the engagement?

Does air have mass? Does gas have mass? Turn-and-talk: take all student answers. Do not correct misconceptions at this time.

Dry ice demonstration: Dry ice is CO_2 in solid form, which has noticeable mass. Obtain about 40 grams of dry ice, demonstrate weighing and recording the mass of CO_2 chunk for students, and place the chunk of ice into a large sealed plastic trash bag. The solid CO_2 chunk will sublimate into gas and expand to about 22 liters of gas. Ask students again: Does gas have mass?

In advance, use a heated paper clip or tiny nail to punch three holes of the same size in a vertical line on a two-liter plastic soda bottle. Punch the holes about 4–5 cm apart (starting at about 4–5 cm from the bottom of the bottle), and cover with a vertical strip of duct tape or electrician's tape (about 18 cm long) so as to temporarily seal all three holes. Then fill the bottle with water, and tightly cap. Be careful not to squeeze the sides of the bottle. Allow students to work in pairs (think–pair–share). Ask the students the following question: What is in the bottle? Take all answers. Then explain to students that the tape is covering three tiny holes.

Students predict what may happen as the tape is removed from each hole (start with the top hole) and hypothesize about the cause of the observed results as one piece of tape is removed at a time (liquid flow under air pressure, balanced and unbalanced forces, surface tension of water). Students then draw or describe in their notebooks what they predict will happen. Then they should draw or describe what actually happened.

EXPLORE

- Describe the hands-on activities that will provide experience of the phenomenon.
- List "big idea" conceptual questions you will use to encourage and focus students' exploration and to allow the students to test their ideas.

Balloon Thermometer

Teacher note: Make sure that students do not have allergies to latex. Students will be exploring how a change in temperature will affect the volume of a balloon when the pressure remains constant. They will first find the volume of a balloon in room-temperature water using the displacement method of volume measuring. Students will mark the water level in a beaker of water with and without the balloon in the water. The difference in water levels will be the volume of the balloon. They will then perform at least two more trials, one with colder water and one with warmer water. List the "big idea" conceptual questions that the teacher will ask to focus the student exploration: Does the temperature of water directly correlate to volume?

Procedure

1. Blow up a balloon to about the size of a baseball.

2. Fill a 1000 ml beaker about half full with tap water and mark the outside of the beaker at the waterline using a whiteboard marker.

3. Submerge the balloon in the beaker and hold it under water for three minutes using tongs to keep the balloon in the water.

4. Take the temperature of the water using the digital thermometer in units of °C and record the value in the data table.

5. Mark the new waterline on the side of the beaker and remove the balloon.

6. Fill graduated cylinder with water and record the water volume as the "initial volume."

7. Pour water from the graduated cylinder into the beaker until you reach the second waterline mark (you may have to refill this several more times, so make sure to keep track of how much water is being added to reach the mark). Once you have reached this mark, record the new volume of water in the graduated cylinder as the "final volume."

8. Empty the beaker and remove the waterline marks.

9. Fill beaker about one quarter full with ice. Add water till about half full.

10. Repeat Steps 3 through 9 for cold water bath.

11. Fill the beaker about half full with water and heat at Setting 6 for about five minutes.

12. Repeat Steps 3 through 9 for hot water bath.

13. Make a graph of the results.

See Table 9.1 for a data collection chart for this activity.

Table 9.1. Data collection chart for Explore activity for Lesson PS1.A

Data Collected	Ice Bath	Room Temperature	Warm Water Bath
Temperature			
Initial volume			
Final volume			
Volume of the balloon (final–initial)			

EXPLAIN

- Student explanations should precede your explanations or introduction of terms. What questions or techniques will you use to help students connect their exploration to the concept under examination?

- List higher-order thinking questions that you will use to solicit *student* explanations and to help students construct and justify their explanations

Ask students the following questions:

- Did any additional air enter the balloon while the tests were being conducted?
- What must be happening to the air in the balloon?
- What is pushing on the inside of the balloon? Is this considered a force?
- Is the air in the balloon moving? Why or why not?
- If the balloon was not there, what would happen to the air?
- What happens when you pop the balloon? Is that different from when you blow up a balloon and then release the air from it?
- What is a common word used for moving air?
- What is the relationship between temperature and volume?

Mathematics Connection: Using their graphs, have students estimate the volume of their balloon given that the temperature was 97°C.

In their science notebooks, students should draw what is happening with the air inside the balloon and then draw what happens to the air when the balloon is popped. The teacher

circulates the room to ensure students have a common understanding that air is made of molecules too small to be seen.

ELABORATE

- Describe how students will develop a more sophisticated understanding of the concept.
- What vocabulary will be introduced, and how will it connect to students' observations?
- How is this knowledge applied in our daily lives?

Now that students understand that air is made of molecules too small to see and that these molecules, when moving, cause a force, it is time to find out how much force can be harnessed from moving air. Another term for moving air is *wind*.

StEMT*ify*

Science: What real-world issues and problems exist in the area of clean energy?

Engineering: Is the STEM lesson guided by the engineering design process?

- Students should be immersed in hands-on inquiry and open-ended exploration.
- Students should be involved in productive teamwork.

Mathematics: Does the STEM lesson apply rigorous mathematics and science content that the students are learning?

Technology: Does the STEM lesson allow for multiple right answers through prototype development, testing, and improvement of design? Encourage students to monitor and evaluate their progress, change course if necessary, and ask, "Does this make sense?"

Present the students with the engineering design challenge:

StEMT*ified* Question: How do windmills work? How does windmill technology support the claim that matter is everywhere and can be too small to be seen?

Step 1: Ask—Practice of asking questions (science) and defining problems (engineering)

What is the problem to solve? What needs to be designed, and who is it for? What is the goal?

Explain that students must develop their own working windmill from everyday items and that the windmill must be able to withstand a medium-speed fan for one minute while winding a string to lift a small object such as a tea bag. (Note: As an extra challenge, test the windmill's ability to lift heavier objects such as coins or washers.)

Step 2: Research and Design—Practice of planning and carrying out investigations

What are the project requirements and limitations?

Students will be given a "budget" from which they will need to purchase materials you provide. Assign a cost for each item that will result in the average team being able to purchase at least 30 material parts.

Step 3: Plan—Practice of constructing explanations and designing solutions

Students meet and develop a plan for their windmill. They agree on the materials they will need, write or draw their plan, and then present their plan to the class.

Step 4: Create—Practice of developing and using models

Student groups next execute their plans. Student teams may request to exchange materials or to order more materials from the teacher. They may also trade unlimited materials with other teams to develop their ideal parts list. They will need to determine the "cost" of their design. The cost of the design will be factored into the efficiency of the model.

Step 5: Test and improve—Practice of obtaining, evaluating, and communicating information

Teams will test their windmills with the fan or hair dryer set up. (Note: you may wish to make the fan available during the building phase so they can test their windmill during the building phase prior to the classroom test.) Each team will test their windmill using a classroom fan or hairdryer. Each windmill should be tested using the same wind speed—medium—at a distance of three feet (one meter). Students need to make sure that their windmill can operate for a

minute at this speed while winding a light object up with a string. Students should watch the tests of the other teams and observe how their different designs worked.

Look-Fors

Students should engage in the following practices:

- *Practice of constructing explanations and designing solutions.* Students will design, plan and carry out investigations to collect and organize data (SEP 3; MP 1).

- *Practice of developing and using models.* Students will obtain, evaluate, and communicate information by constructing explanations and designing solutions (SEP 8; MP 3). Students will develop and use models by designing a windmill prototype for simulations SEP 2; MP 4).

EVALUATE

- How will students demonstrate that they have achieved the lesson objective?

- Evaluation should be embedded throughout the lesson as well as at the end of the lesson.

Teams present their windmills to the class and answer the following questions: Did you succeed in creating a windmill that could lift an object the required height? If not, why did it fail? In what ways did you revise your original design? Why?

Summative Assessment: Write about it!

Using the claims, describe how the windmill prototype works and how it demonstrates that matter (air) may be made up of particles that are too small to be seen but still have mass. Provide evidence for this claim using information from the Explore or StEMT*ify* portion of this lesson. Provide a statement that explains how your evidence provides support for your claim.

Aircraft Catapult

Motion and Stability: Forces and Motion (PS2.A)

Teacher Brief

Ramps and sliders are common tools to help students understand the fundamentals of Newton's laws. The Explore activity included is a typical Newton's Second Law activity, but other concepts, such as inertia and friction, can be studied with this same setup. Depending on grade level, state standards, and ability, the teacher's choice of Explore activities preceding the engineering design challenge may be slightly different. When we have done this lesson with students we enjoyed the discussions of the practice of science that can follow (e.g., independent, dependent, and control variables).

The engineering design challenge is a culmination of many attempted activities that resulted in varied results. Our first attempt controlled the materials and construction methods too much. Although it focused student attention more on Newton's laws which control the actions, it eliminated student interactions with the engineering design process. Once we allowed more student independence, the students became more empowered, although, it did require a bit more time for construction. However, during the construction time, opportunities for discussions between the teacher and each group proved invaluable. We were able to ask students why a design failed and brought that failure into a discussion about forces such as friction and balanced and unbalanced forces. Our advice is to toss out random materials, whether or not you think they are usable. Do not limit your students. Our young scientists often surprise us with designs that are more creative than what we had anticipated. Once we realized the errors in our ways with material constraints, the students became more empowered and as long as we made sure we circulated amongst groups and asked questions, the focus on Newton's laws was not lost. In fact, the opportunities for discussion were enhanced. As an example, as student designs fail, ask how the failure was related to concepts such as friction and balanced and unbalanced forces.

By the end of grade 5. Each force acts on one particular object and has both a strength and a direction. An object at rest typically has multiple forces acting on it, but they add up to give zero net force on the object. The patterns of an object's motion in various situations can be observed and measured; when past motion exhibits a regular pattern, future motion can be predicted from it.

Framework **Question:** How can one predict an object's continued motion, changes in motion, or stability?

StEMT*ified* Question: How can an aircraft carrier catapult be designed so that it will launch aircraft of different masses at an appropriate (safe) speed?

Dimension 1: Practices (What should students be doing?)

- Developing and using models
- Using mathematics, information and computer technology, and computational thinking
- Planning and carrying out investigations
- Constructing explanations and designing solutions

Dimension 2: Crosscutting Concepts

- *Cause and effect.* Cause-and-effect relationships are routinely identified, tested, and used to explain change. Students will plan and conduct an investigation to provide evidence of the effects of balanced and unbalanced forces on the motion of an object.

- *Systems and system models.* Defining the system under study—specifying its boundaries and making explicit a model of that system—provides tools for understanding and testing ideas that are applicable throughout science and engineering. Models can be used to represent systems and their interactions, and to show that energy and matter flows within systems.

Dimension 3: Disciplinary Core Idea

Interactions of an object with another object can be explained and predicted using the concept of forces, which can cause a change in motion of one or both of the interacting objects. An individual force acts on one particular object and is described by its strength and direction. The strengths of forces can be measured and their values compared (*Framework;* NRC 2012).

Instructional Duration: Four to five 45-minute class periods

Materials

Per class: overhead projector or document camera; balance; masses for balance

Per group: masking tape, meter stick or measuring tape, ruler with groove down the center, or commercial ramp and slider kit, marbles, ½ paper cup (cut lengthwise or a small milk carton with an open end), calculator, sheet of graph paper, 12 pennies

Per student: science notebook and pencil, safety glasses or goggles

Safety Notes

1. Personal protective equipment (safety glasses or goggles) should be worn during the set-up, hands-on, and take down segments of the activity and demonstration.

2. Immediately pick up any marbles off the floor.

3. Wash hands with soap and water upon completing this activity.

Guiding Question: What are some effects of balanced and unbalanced forces on the motion of an object?

Misconceptions

- Friction only slows objects down.
- The natural state of objects is at rest; When an object is at rest, no forces are acting upon it.
- Forces are caused by moving objects.
- When forces are balanced, no motion can occur.

ENGAGE

- How will you capture students' interest and reveal misconceptions?
- What kinds of questions should the students ask themselves after the engagement?

Teacher Questions: Why is it important for a football linebacker to be big compared to a quarterback or a wide receiver? Why do some people think bigger cars are safer than smaller cars? What is a catapult and how does it work?

Demonstration through Tug-O-War: What happens when the forces are balanced? What happens if the forces are unbalanced? What causes the forces to be unbalanced? Explain that unbalanced forces cause a change in motion.

EXPLORE

- Describe the hands-on activities that will provide experience of the phenomenon.

- List "big idea" conceptual questions you will use to encourage and focus students' exploration and to allow the students to test their ideas.

Before starting the experiment, establish these guidelines:

- All experiments should take place in the area that you designate for each group.

- Make sure the sliders are placed even with the end of the ramp for each trial.

- The spheres should be released down the ramp – not pushed.

- Remember to measure the mass of the spheres AND the mass placed in the slider.

- Recheck the set up before starting each trial.

(*Note:* If you are using rulers, marbles, and cups instead of commercial ramps and sliders, have students set the 4 cm mark of the ruler on a textbook to form a ramp.

Distribute the materials. Ask students to discuss with their group how they will set up an experiment to test the testable question we established as a class.

Have groups report out their ideas to the class prior to beginning so you can ensure that groups are on the right track. The lab procedure is written as an example of how the experiment can be completed. Allow students to complete their experiments in various ways as long as they are following proper scientific procedures.

- Students should take turns releasing the sphere three times for each mass.

- Each time, students should record the distance the slider travels by measuring in centimeters from the end of the ramp to the leading edge of the slider. Students should copy these recorded data charts into their science notebooks.

- Each test should be repeated for a minimum of three trials with results averaged.

- Have students create a bar graph of their data in their science notebook for each distance the sphere travelled before hitting the slider.

- Collect one group's data and create a bar graph on a transparency or project using the document camera. Display the graph for the class. Discuss the results with the class and ask if they can see any pattern in the data.

- Continue collecting data on the distance slider moves by moving the sphere to a higher or lower distance on the ramp to show the increase or decrease of forces on the object (slider).

See Table 9.2 for a data collection chart.

Table 9.2. Data collection chart for Explore activity for Lesson PS2.A

Number	Mass	Trial (mm)	Trial 2 (mm)	Trial 3 (mm)	Trial 4 (mm)	Trial 5 (mm)	Trial 6 (mm)
1 mass							
2 masses							
4 masses							
6 masses							
8 masses							

Data analysis: Plot these data points on a line graph to compare the distance the cup moved with the amount of mass sitting on top of the cup. Remember to plot the independent variable on the *x*-axis and the dependent variable on the *y*-axis. Make sure the graph has a descriptive title. Also include axis labs and units.

- What is the title of the graph?

- What is the label on the *x*-axis (w/units)? What is the label on the *y*-axis (w/units)?

- Place a line of best fit through your data points. Describe what this line looks like.

- By analyzing the data from the data table and the graph, what can be said about what happens when mass is added to an object as it relates to its motion?

Much of the reason for graphing data is that an infinite amount of data is impossible to test. By having a graph, use the line of best fit to predict an outcome for untested data. This *interpolation* allows us to "fill in the blanks." Students make a prediction based upon the graph as to the effect (in centimeters) of adding 3 and 4 masses to the cup. Students should extend the line of best fit out past the last data point to allow for *extrapolation,* or predict for data that is beyond the range of tests.

- What would happen if 10 masses were added?

- Is there a point where the maximum distance is reached regardless of the amount of mass used? Support your answer using scientific terms.

- As a scientist, which would you feel safer doing, extrapolating or interpolating? Why?

- Why would the masses fall off the top of the cup if they were not held on by some means?

Error analysis: What are some potential errors that may occur during this lab and what are the potential sources of these errors?

EXPLAIN

- Student explanations should precede your explanations or introduction of terms. What questions or techniques will you use to help students connect their exploration to the concept under examination?

- List higher-order thinking questions that you will use to solicit *student* explanations and to help students construct and justify their explanations.

Teacher questions for small group discussion:

- What mass of the slider made the slider travel farther? Why? *[Answer: An object's acceleration depends on the mass and the force applied. Because we are keeping the force, (the push of the marble) the same while the mass of the slider changes, there will be a difference in acceleration.]*

- What mass of the slider caused no motion of the slider? Why? *[Answer: If the force, the marble being rolled, and the mass of the slider were balanced, there would be no acceleration, or motion.]*

- From where did the energy come to move the sphere and the slider? *[Answer: The sphere is sitting at rest at a level on the ramp; it is not moving, but it has stored energy, called gravitational potential energy due to its position. When the sphere was released, potential energy was converted to kinetic energy. When the sphere hit the slider, energy was transferred from the sphere to the slider and the slider moved.]*

- Why did we repeat the experiment? *[Answer: To ensure more accurate results.]*
- Was our question testable? Did it provide data for use in answering the key question?

Student task: Using the evidence collected, have students write a brief paragraph answering the guiding question. What are some effects of balanced and unbalanced forces on the motion of an object? Explain how your data support or refute your hypothesis.

ELABORATE

- Describe how students will develop a more sophisticated understanding of the concept.
- What vocabulary will be introduced, and how will it connect to students' observations?
- How is this knowledge applied in our daily lives?

Have students repeat the investigation. Students should compare their results and discuss similarities and differences in results and the reasons for these similarities and differences. How does the students' knowledge of Newton's Third Law assist in determining how well the design solution meets the criteria and constraints? How are the technologies used in the design affected by the constraints and limitations of the problem?

StEMTify

Science: What real-world issues and problems exist in the area of transportation and use of mass and forces?

Engineering: Is the STEM lesson guided by the engineering design process?

- Students should be immersed in hands-on inquiry and open-ended exploration.
- Students should be involved in productive teamwork.

Mathematics: Does the STEM lesson apply rigorous mathematics and science content that the students are learning?

Technology: Does the STEM lesson allow for multiple right answers through prototype development, testing, and improvement of design? Encourage students to monitor and evaluate their progress, change course if necessary, and ask, "Does this make sense?"

StEMT*ified* Question: How can an aircraft carrier's catapult be designed so that it will launch aircraft of different masses at an appropriate (safe) speed?

Student materials per group for StEMT activity: one four-foot-long table, two small bull (binder) clips, one Ziploc baggie to hold the sand (force applying falling mass), four small Ziploc baggies (projectile), sturdy string (several meters), two cups of sand, triple-beam balance or electronic scale, one spoon to scoop the sand into and out of the baggie, plastic cups, paper cups, cardboard, or other material students can use to build the projectile holder, smooth piece of wood or laminate flooring to act as ramp (because our projectiles do not have wings, they cannot produce lift, so lift will be supplied by a shallow ramp to achieve some degree of vertical launch capacity). *Note: This challenge can be done without a ramp.*

Safety Notes

1. Personal protective equipment (safety glasses or goggles) should be worn during the set-up, hands-on, and take down segments of the activity and demonstration.

2. Make sure all fragile items are removed from the potential projectile trajectory.

3. Wash hands with soap and water upon completing this activity.

Step 1: Ask—Practice of asking questions (science) and defining problems (engineering)

What is the problem to solve? What needs to be designed, and who is it for? What are the project requirements and limitations? What is the goal?

Students will need to build and test a model of an aircraft carrier launcher. The students will be provided initially with three baggies filled with sand of varying amounts. They will need to determine the force needed to launch these projectiles a given distance, which will represent the launch speed needed to get an aircraft airborne without damaging the plane or the people on board. Students will construct a mathematical model that they can use to predict the force needed to launch other aircraft of different masses. Once the students are comfortable with their model, they will be given a fourth mass to launch. They must use their mathematical model to determine the force (amount of sand added to the falling baggie) to launch the plane.

Step 2: Research and design—Practice of planning and carrying out investigations

Students collaborate to brainstorm ideas and develop as many solutions as possible. Students will be given the materials to construct the aircraft launcher.

Explain that the small baggie of sand will represent the aircraft's mass. Not all aircraft launched from the deck of the carrier will have the same mass. Some planes are small and fast; others are large and loaded with fuel or supplies. The catapult must be set to accommodate the plane's weight. Too fast a launch will damage the plane. Too soft a launch will not get the plane airborne.

Step 3: Plan—Practice of constructing explanations and designing solutions

Students will compare the best ideas, select one solution and make a plan to investigate it.

Step 4: Create—Practice of developing and using models

Students will build a prototype.

1. Provide students with a short ramp to set at the end of the table. It works best if this is taped in places at the bottom. It should be held up on one end by textbooks or small blocks of wood.

2. Students will first need to construct the projectile holder. This will be where the projectile sits as it is pulled down the ramp. Provide students with material options. Once students choose their materials, they cannot substitute them for other materials without the instructor's permission.

3. Students will attach the bull clip to each end of the string. (Let the students determine how long the string needs to be.)

4. The string will hang off the edge of the table so the larger baggie of sand can fall, supplying the force for the launch.

5. Students will fill three baggies (non-Ziploc) with sand to act as the projectiles. The first projectile should have a mass of 10 grams. The second projectile will have a mass of 20 grams and the third projectile 40 grams. Once the sand is measured into the baggie, twist the top of the baggie closed and secure it with tape. It works best if the baggie is trimmed just above the tape to minimize its size.

6. Place one strip of tape on the floor about three meters from the launch end of the table. Place the second strip approximately 0.5 meters farther from the table than the first.

7. Students will select a mass of sand for the large (force-supplying) baggie. Students will measure the mass of sand needed for their first test and add it to the small baggie (the projectile).

8. Students will connect the force-supplying baggie of sand to the string that will be hung off the edge of the table and will pull the launcher cup up the ramp.

9. The added sand will need to supply the correct amount of force to launch the projectile the given distance (i.e., between the two lines). This distance equates to the correct launch speed.

10. Once students have figured out how much mass is needed to launch the first projectile, they will repeat the procedure to launch the next two projectiles of different masses.

11. Students will take their data from the successful launches and record them in a graph.

12. Students will determine a line of best fit.

Step 5: Test and improve—Practice of obtaining, evaluating, and communicating information

Does the prototype work, and does it solve the need? Students communicate the results and get feedback and analyze the results and talk about what works, what does not, and what could be improved.

- The students will now be supplied with a fourth mass (50 grams of sand).

- Using the line of best fit on their graph, students can now make a prediction as to the amount of force needed (mass of the hanging baggie) to launch the fourth projectile the correct distance.

- For this test, groups will get three tries to get their projectile within the landing zone indicating a successful launch.

See Table 9.3 for a data collection chart.

EVALUATE

- How will students demonstrate that they have achieved the lesson objective?

- Evaluation should be embedded throughout the lesson as well as at the end of the lesson.

Presentation

Students should prepare a class presentation about the benefits of their design (along with the noted limitations) and the effectiveness of their mathematical model in making predictions using the *Claims, Evidence, Reasoning* framework. Students should use appropriate vocabulary learned throughout the lesson (e.g., force, mass, inertia) to explain their catapult design.

Table 9.3. Data collection chart for StEMT activity for Lesson PS2.A

Projectile 1 Mass (g)	Mass Supplying Force (g)	Distance Traveled (cm)	Successful (Yes or No)	Amount of Change Needed? (More or Less Force)
Trial 1				
Trial 2				
Trial 3				
Projectile 2 Mass (g)	**Mass Supplying Force (g)**	**Distance Traveled (cm)**	**Successful (Yes or No)**	**Amount of Change Needed? (More or Less Force)**
Trial 1				
Trial 2				
Trial 3				
Projectile 3 Mass (g)	**Mass Supplying Force (g)**	**Distance Traveled (cm)**	**Successful (Yes or No)**	**Amount of Change Needed? (More or Less Force)**
Trial 1				
Trial 2				
Trial 3				

Writing Assignment

Students can prepare a formal report for a prospective client who is interested in purchasing the new launch mechanism and mathematical model to use on new military ships the client has been contracted to build.

Keepin' It Cool!

Physical Sciences: Energy—Conservation of Energy and Energy Transfer (PS3.B)

Teacher Brief

We have fond memories of this lesson. It was a version of this lesson that set us on the path to write this book. Our state's STEM coordinator asked if they could visit our district and make a video of a STEM lesson. A few phone calls later and we adopted a fourth-grade classroom for a few days. The difficult part was developing a STEM lesson worthy of being recorded and shared with other teachers from around the state.

The lesson itself focuses on heat transfer and how materials have characteristics that can increase or decrease the amount of heat flow through them. The students were going to test different materials to see which material was the best insulator. Using some simple materials (foam board, cardboard, card stock and aluminum foil, etc.), students tested the ability of insulators to prevent ice cubes from melting.

The Engage portion starts with a simple discussion to review prior knowledge. The Explore section starts the investigation, with students investigating how heat is transferred and what types of materials can limit the heat transfer. If you live in a sunny warm climate, the activity can be set up outside. If not, a lamp of some sort may be necessary to provide the focused heat. Note— making observations every three minutes helps keep this lesson moving. Do not miss the opportunity to discuss with groups the concepts that involve the practice of science, including making observations, recording data, and drawing conclusions. Graphing can, and probably should, be an integral piece of this initial investigation of the materials.

The engineering design challenge uses the knowledge gained within the Explore section to develop a solution to a problem. We have provided some parameters as a guide, but you may wish to modify them to meet specific needs. Essentially, students are constructing a container capable of keeping something frozen for as long as possible. There are many opportunities throughout this activity to bring in data analysis and graphing. When we did this lesson with fourth graders at one of our most challenging schools, they were graphing.

Another option to consider for this engineering challenge is to substitute technology for the simple model we have shared. Instead of melting some volume of ice, you can use a metal temperature probe available from many companies (e.g., Pasco, Vernier). Simply place the metal probe into ice water to cool, and the metal rod acts as a thermal mass. Instead of using a larger box, a box just big enough to contain the length of the metal probe should be designed. This goes really quickly and requires students to make temperature observations every 30 seconds. The data produced are engaging. If you can get your hands on probeware, by all means, do so.

We also used this probeware option with fourth graders who had no experiences with probeware until that day. We thought we would need to provide step-by-step instructions, but we were quickly proven wrong. Students are not intimidated by technology, regardless of their socio-economic status. The students we test our lessons with come from high-poverty schools. Having used both versions, either will work well to help students understand heat transfer and support the knowledge of conductors and insulators.

By the end of grade 5. Energy is present whenever there are moving objects, sound, light, or heat. When objects collide, energy can be transferred from one object to another, thereby changing their motion. In such collisions, some energy is typically also transferred to the surrounding air; as a result, the air gets heated and sound is produced. Light also transfers energy from place to place. For example, energy radiated from the Sun is transferred to Earth by light. When this light is absorbed, it warms Earth's land, air, and water and facilitates plant growth. Energy can also be transferred from place to place by electric currents, which can then be used locally to produce motion, sound, heat, or light. The currents may have been produced to begin with by transforming the energy of motion into electrical energy (e.g., moving water driving a spinning turbine, which generates electric currents. *Framework*; NRC 2012).

Framework Questions: What is meant by conservation of energy? How is energy transferred between objects or systems?

StEMT*ified* Question: How can we keep perishable items cold enough to stay frozen?

Dimension 1: Practices (What should students be doing?)

- Planning and carrying out investigations
- Constructing explanations and designing solutions
- Obtaining, evaluating, and communicating information

Dimension 2: Crosscutting Concepts

- *Energy and matter: Flows, cycles, and conservation.* Tracking fluxes of energy and matter into, out of, and within systems helps one understand the systems' possibilities and limitations. Energy can be transferred in various ways and between objects.

Dimension 3: Disciplinary Core Idea

The total change of energy in any system is always equal to the total energy transferred into or out of the system. This is called conservation of energy. Energy cannot be created or destroyed, but it can be transported from one place to another and transferred between systems. Energy can also be transferred from place to place by electric currents. Heating is another process for transferring energy. Heat transfer occurs when two objects or systems are at different temperatures. Energy moves out of higher temperature objects and into lower temperature ones, cooling the former and heating the latter. (*Framework*; NRC 2012).

Guiding Question: How does the amount of heat energy affect water?

StEMT*ified* Question: How can we keep perishable items cold enough to stay frozen?

Apply scientific ideas to design, test, and refine a device that converts energy from one form to another. This, of course, is not the only question that would work or the only way to approach this lesson. You could also use the original question during the first section of the lesson. The second, problem-based question is introduced after the concept has been mastered and students are ready to apply their knowledge within the Elaborate portion of the lesson.

Misconceptions

- Cold can be transmitted from one object to another.
- A phase change (e.g., ice to water) results in a new substance.

Instructional Duration: Four to five 45-minute class periods

Materials per Group

Ice tray or medicine cups, aluminum foil, bubble wrap, cardboard, foam board (available from craft store), halogen lamp (or other bright light that produces a lot of heat), small funnel, 250 ml beaker, 50 ml graduated cylinder, stopwatch or timer, indirectly vented chemical splash goggles, nonlatex aprons

Safety Notes

1. Personal protective equipment (splash goggles, aprons) should be worn during the setup, hands-on, and takedown segments of the activity and demonstration.

2. Use caution when heating or handling hot substances (hot water, etc.) as they can burn skin.

3. Keep electrical devices (halogen lamp, etc.) away from water sources (shock hazard).

4. Use caution when handling glassware (can shatter and puncture skin).

5. Immediately wipe up any spilled water on the floor (slip-and-fall hazard).

6. Wash hands with soap and water upon completing this activity.

For a basic overview of physical science, teachers may use the OER CK–12 Standards-Aligned FlexBook textbook *Physical Science* (*www.ck12.org/teacher*).

ENGAGE

- How will you capture students' interest and reveal misconceptions?
- What kinds of questions should the students ask themselves after the engagement?

Ask the students to make a list of places they can find water. Take all answers. Do NOT correct misconceptions at this time. If students provide examples of ice or vapor, ask students if these examples are still water. Introduce the term *state* and the phrase *states of matter*. Create class definitions of solids, liquids, and gases. Give each student an ice cube in a plastic zipper-type bag. Ask students to see if they can melt the ice cube without taking it out of the bag. Create a data table recording the methods attempted and the amount of time it took to completely melt the ice cube. What is happening? What is causing the ice to melt? What are some other ways we could melt the ice?

EXPLORE

- Describe the hands-on activities that will provide experience of the phenomenon.

- List "big idea" conceptual questions you will use to encourage and focus students' exploration and to allow the students to test their ideas.

Part 1: Provide each group with a tray of frozen ice cubes or 12 individual 20 ml medicine cups containing equal amounts of ice (two for each type of insulating material being tested). Groups will use either the ice cube trays or the frozen medicine cups to make observations. Students will record their observations in their notebooks using the Table 9.4 as a guide.

Students will explain what happened under each type of material using the terms *solid, liquid,* or *gas* as well as m*elting, freezing, boiling, evaporation,* and *condensation*. Each diagram will NOT have all of these terms. However, students MUST record data using each of the words relating the state of matter that the water is in with the correct term. For example: students draw a picture of an ice cube, labeling it solid and frozen.

Cover the ice cube trays or frozen medicine cups containing ice with equally sized pieces of the following material: bubble wrap, cardboard, aluminum foil, black-and-white construction paper and foam board. Leave samples of ice uncovered as a control group. Place the samples under a halogen lamp or in the direct sunlight. Make observations of the ice under the material every 3 minutes. *Note: Time may vary with the amount of heat being applied.* See Table 9.4 for a data collection chart for this activity.

Table 9.4. Data collection chart for Lesson PS3.B

Trial	Three minutes	Six minutes	Nine minutes
Ice uncovered			
Cardboard			
Bubble wrap			
Aluminum foil			
White construction paper			
Black construction paper			
Foam board			

Part 2: Emphasize to students that the increased heat energy not only caused the ice to melt quickly but that some of it may have turned into a gas and escaped into the air. If we added heat to change the ice to water and the water to a gas, what could we do now to make the gas turn back into a liquid? How can we turn the liquid back into a solid?

Procedure

Provide each group with a Styrofoam cup containing warm water. Have students cover the top of the cup with plastic wrap and secure it with a rubber band. Provide students with one ice cube that they will place on top of the plastic wrap. Students should draw a diagram and label it using the terms *solid*, *liquid*, or *gas,* as well as *melting, freezing, boiling, evaporation,* and *condensation*. Each diagram will NOT have all of these terms. However, students MUST record data using each of the words relating the state of matter that the water is in with the correct term. For example: students draw a picture of steam labeling it gas and evaporating. Have students answer the following questions in their science notebooks:

- When did we have solids? When did we have liquids? When did we have gases?

- In what ways did the material covering the ice affect its rate of melting?

- When you change the way something LOOKS, but it is still the same object, that is a physical change. When we heated and cooled the water, what kind of changes were those? How do you know?

- What is ice made of? Is ice still water?

- When we heat the ice, what happens? Is this still water?

- When we looked at the lid to open the pan and saw the **condensation**, was this water?

EXPLAIN

- Student explanations should precede your explanations or introduction of terms. What questions or techniques will you use to help students connect their exploration to the concept under examination?

- List higher-order thinking questions that you will use to solicit *student* explanations and to help students construct and justify their explanations.

Have students rank their materials from best to worst in terms of preventing the ice from melting on a piece of chart paper or on a whiteboard. Give groups a chance to compare their results via a gallery walk. Ask students the following questions for small group discussions:

- What can we say about melting ice from the data displayed?

- What seems to be the best material at preventing the ice from melting?

- What do the best materials have in common?

- Did anyone use the same material but get different results? Why do you think this is?

- Why do you think scientists share their results with each other?

Have students create a flow map showing the words *boiling, melting, condensation, evaporation*, and *freezing*. The arrows should be labeled either "add heat energy" or "take away heat energy," depending on the order of the words. Have students create a flip book (piece of construction paper folded long, with the front part cut into five parts. Students can draw condensation, freezing, boiling, evaporation, and melting on the flaps and write an explanation of what happens to water during each of those under each flap.

ELABORATE

- Describe how students will develop a more sophisticated understanding of the concept.

- What vocabulary will be introduced, and how will it connect to students' observations?

- How is this knowledge applied in our daily lives?

Original Elaborate for this lesson: Ask students to estimate how long it would take the ice cube to melt at room temperature, in the sunlight, in a cooler, or in a refrigerator. Invite other suggestions and explore these questions through experimentation as time allows.

StEMT*ify*

Science: What real-world issues and problems exist in the area of food safety?

Engineering: Is the STEM lesson guided by the engineering design process?

- Students should be immersed in hands-on inquiry and open-ended exploration.
- Students should be involved in productive teamwork.

Mathematics: Does the STEM lesson apply rigorous mathematics and science content that the students are learning?

Technology: Does the STEM lesson allow for multiple right answers through prototype development, testing, and improvement of design? Encourage students to monitor and evaluate their progress, change course if necessary, and ask, "Does this make sense?"

StEMT*ified* Question: How can we keep perishable items cold enough to stay frozen?

Step 1: Ask—Practice of asking questions (science) and defining problems (engineering)

What is the problem to solve? What needs to be designed, and who is it for? What is the goal?

In many parts of the world, refrigeration is not available. It can also take many days to travel to remote parts of the globe. Many foods must be kept frozen until eaten. To transport frozen food to these parts of the world, it is critical to have containers that limit the amount of heat entering the package. In poorer regions of the world, it is also important to keep the cost down. Given these constraints, how can a low-cost but efficient package be designed?

Step 2: Research and design—Practice of planning and carrying out investigations

Students collaborate to brainstorm ideas and develop as many solutions as possible.

Explain to students they have just conducted an investigation with ice being melted at different rates because the material between the ice and the heat source was acting as an insulator. Insulators restrict the amount of energy (i.e., heat) that can pass through them. From their observations, students will notice that some materials are better insulators than others. The best insulator kept the ice from melting for the longest time. The worst insulator allowed the ice to melt more quickly.

Have students describe their findings in their science notebooks for the following questions:

- Which material made the biggest difference in the time it takes ice to melt?

- Which material provided the least insulation and allowed the ice to melt quickly?

- When ice melts it turns to water. Are there some of the materials that might not react well to being wet?

- What are some ways to prevent some of these materials from becoming wet from the melting ice?

- Do any of the materials tested have the ability to prevent other materials from becoming wet, which may affect the strength of the material if it were used to construct a container to hold ice?

Teacher note: You will need to have medicine cups filled with 15 ml of water and frozen overnight for this engineering design challenge.

Students will be given an engineering design challenge where they need to design, build and test a frozen food container.

Parameters: The package must be constructed of foam board to a size of 10 x 10 x 10 cm. The container needs to have a side door that will open so that the ice samples can be placed into the box. Each side can have no more than 3 layers, with each layer being constructed from a different material. Duct tape or masking tape will be used to hold the sides together. In the bottom of the container there should be a small hole so that the stem of a small funnel can be placed through it. Students will be given materials to use to construct their 10 x 10 x 10 cm frozen food package: cardboard, bubble wrap, aluminum foil, black-and-white colored construction paper, and foam board.

Step 3: Plan—Practice of constructing explanations and designing solutions

Students will compare the best ideas, select one solution, and make a plan to investigate it.

Step 4: Create—Practice of developing and using models

Students will build a prototype.

The foam board should be used to make the square container. The other materials can either be placed on the outside of the container or on the inside. During the engineering design challenge, only the materials other than the foam board box can be changed.

1. Place a 500 ml beaker under the box to catch the water and to support the box.

2. Place the medicine cup containing the ice cube into the funnel and insert the funnel and ice into the box.

3. Seal the box's side door with a small piece of masking tape.

4. Place the box on top of the wire screen so that it completely covers the medicine cup.

5. As soon as the apparatus is set up, turn on the heat source and start the timer.

Step 5: Test and improve—Practice of obtaining, evaluating, and communicating information

Does the prototype work, and does it solve the need? Students communicate the results and get feedback. Students then talk about what works, what doesn't, and what could be improved.

1. Observe carefully as the ice melts and turns to a liquid. The liquid will drain into the beaker. Stop the timer when no more water is dripping from the container and into the beaker or the amount of known water has been collected in the beaker below the container.

2. Continue using the engineering design process to improve your container.

Look-Fors

Students should engage in the following practices:

- *Practice of planning and carrying out investigations.* Students will be actively engaged and work cooperatively in small groups to complete investigations, test solutions to problems, and draw conclusions. Use rational and logical thought processes, and effective communication skills (writing, speaking and listening; SEP 7, SEP 8; MP 3).

- *Practice of constructing explanations and designing solutions.* Students will design, plan and carry out investigations to collect and organize data (SEP 3; MP 1).

- *Practice of obtaining, evaluating, and communicating information.* Students will analyze and interpret data to draw conclusions and apply understandings to new situations (SEP 4; MP 5). Students will apply scientific vocabulary after exploring a scientific concept (SEP 6; MP 7).

EVALUATE

- How will students demonstrate that they have achieved the lesson objective?

- Evaluation should be embedded throughout the lesson as well as at the end of the lesson.

Formative Assessment

Occurs continuously throughout the lesson and engineering design challenge.

- Do students understand solids, liquids, and gases?

- Do students understand physical changes?

- Do students understand that throughout each stage, the water was still water?

- Can students explain melting, condensation, evaporation, boiling, and freezing of water?

Summative Assessment: Claim, Evidence, Reasoning—Argue your case!

Upon completion of their design process, students will make a claim about the best materials used to construct a model of a refrigeration container. They will use the evidence collected during the trials to provide a case as to why their design is appropriate. Throughout the claims evidence reasoning process, students should be using the terms correlated with this activity such as *melting, condensation, evaporation, boiling,* and *freezing.*

The Pitch in the Wave

Physical Sciences PS4: Waves and Their Applications in Technologies for Information Transfer—Wave Properties (PS4.A)

Teacher Brief

For most concepts, it is impossible to ensure deep understanding with just one lesson. This concept is a good example of that, as the topic of waves could take many weeks to cover to depth of mastery. For our demonstration lesson, we have focused on a portion of the larger concept: pitch. The concept seems simple, but our students struggle, especially when the perspective of what is vibrating is changed.

One of the most common activities for demonstrating pitch is to have students hang a plastic ruler over the edge of their desk and have their partner pluck it. The students are supposed to notice (see) the change in speed as they move the ruler farther out over the edge. But what are the students hearing, other than the ruler slapping against the desk? The sound made by the motion of the ruler is more noise than sound, making it difficult for the students to make the connection to pitch. The lesson we chose to start with does not include the plastic ruler, for the reasons noted. If you want to use the ruler in an Engage activity, note that it is a more powerful visual connection to pitch than an audible connection to pitch.

Our lesson starts off with glass bottles filled with different amounts of water. Make sure you assist students with recording the distance above the water to the rim of the glass bottle. Resist the urge to explain everything there is to know about this characteristic of sound in order to enhance the anticipation for the rest of the lesson. During the Explore section, we chose to use rubber bands because of their ability to demonstrate pitch and the clear sound they produce. If your classroom acoustics make it sound louder, you might consider having students do this Explore over a small box, or even a small cup, which will help students hear the pitch changes. With our lesson, we chose a method to decrease the number of rubber bands flying across the room or being popped against their partner's hands. The concept is simple enough: students need to shorten the length of the rubber band that is allowed to vibrate. Another method might be to wrap the rubber band around a shoebox or other small box and have students pinch the rubber band at specific locations along its length. Whichever method you use will accomplish the same goal.

Students have seen a rubber band vibrating and have also seen sound being produced in air with the glass bottle. Now it is time for them to start making their own instruments capable of matching certain pitches. The matching pitch concept came about to make sure students make the connection between length and pitch in various scenarios. It is not critical how the pitches are chosen: you can ask students to match one pitch, several pitches, or a familiar song.

Prior Knowledge

By the end of grade 2. Sound can make matter vibrate, and vibrating matter can make sound (*Framework*; NRC 2012, p. 132).

By the end of grade 5. Waves of the same type can differ in amplitude (height of the wave) and wavelength (spacing between wave peaks). Waves can add or cancel one another as they cross, depending on their relative phase, but they emerge unaffected by each other (*Framework*; NRC 2012 p. 132).

Framework Question: What are the characteristic properties and behaviors of waves?

StEMT*ified* Question: How can a knowledge of wavelength and pitch be used to design musical instruments capable of producing a specific pitch?

Dimension 1: Practices (What should students be doing?)

- Developing and using models
- Planning and carrying out investigations
- Constructing explanations and designing solutions

Dimension 2: Crosscutting Concepts

- *Patterns.* Observed patterns of forms and events guide organization and classification, and they prompt questions about relationships and the factors that influence them. Students will develop a model of waves to describe patterns in terms of amplitude and wavelength and to show that waves can cause objects to move.

Disciplinary Core Idea

A simple wave has a repeating pattern with a specific wavelength, frequency, and amplitude. (*Framework*; NRC 2012, p. 131). See Figure 9.1.

Figure 9.1. Wave properties

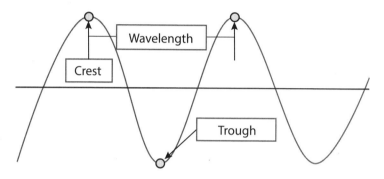

Instructional Duration: Four to five 45-minute class periods

Materials

Teacher Materials (for Engage demo): five identical containers (bottles, jars, or drinking glasses)

   Materials per group: one thin rubber band, three pencils, metal spoon or knife, one small strip of masking tape (approximately 20 cm)

   Materials for Engineering Design Challenge: string or chord, dowel rods, rubber bands, shoe boxes, straws (teachers will demonstrate for students how to cut a reed on the end of the straw), bottles, water (as needed), safety glasses or goggles

Safety Notes

1.   Personal protective equipment (safety glasses or goggles) should be worn during the setup, hands-on, and takedown segments of the activity and demonstration.

2.   Immediately wipe up any spilled water on the floor (slip-and-fall hazard).

3.   Use caution when working with glassware (can fracture and puncture skin).

4.   Wash hands with soap and water upon completing this activity.

   For a basic overview of waves and wave energy, teachers may use the OER CK–12 Standards-Aligned FlexBook textbook *Physical Science* (*www.ck12.org/teacher*).

Misconceptions

- Students think pitch and volume are the same. *[Correct understanding: volume is how loud something is; pitch is how low or high something is based on the speed of vibrations.]*

- Students think the pitch of a tuning fork will lower with the passing of time. *[Correct understanding: the pitch is dependent on the length of the tuning fork. The pitch does not change, even though volume might.]*

- Students think it is the vibrating tuning fork that creates the sound. *[Correct understanding: the sound is created as the air is put in motion because of the movement of the tuning fork.]*

ENGAGE

- How will you capture students' interest and reveal misconceptions?

- What kinds of questions should the students ask themselves after the engagement?

Guiding Question: How does the movement (vibration) of an object affect the sound (pitch)?

Place the five glass bottles, jars, or drinking glasses containing different amounts of water on the table and ask the students if the bottles will make a sound when they are hit. The students should say "Yes!" Ask students to predict which bottle will make the highest sound or pitch. The lowest?

Test the bottles using a metal spoon or knife to cause the vibrations. Help them to hear that the bottle with the smallest amount of water has the highest pitch and that the bottle with the most water has the lowest pitch. With the class, order the bottles from lowest to highest pitch.

Ask: How does the movement (vibration) of an object affect the sound (pitch)? Allow students to their write initial thoughts in their science notebooks.

Cut through a thick rubber band to make one long piece of rubber. Wrap one end of the rubber band around a pencil and tape it so that it is secure. Wrap and tape the other end of the rubber band to another pencil. Make sure the rubber band can stretch about 30 centimeters. Near one end of the rubber band, loop the rubber band once around a third pencil, but do not secure it. One student will hold both secured end of the rubber band so that it is moderately stretched to about 30 centimeters. A second student will hold the unsecured pencil and pluck the rubber band. (See Figure 9.2.)

Figure 9.2. Explore activity

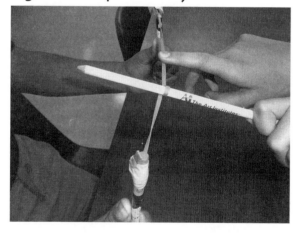

With the rubber band stretched and the pencil near one end, what will happen when the band is plucked on the long side of the stretch band? Was the pitch high or low?

Note: The band must be kept at the same tension (stretched to the same distance).

Students will complete the chart in Table 9.5 (p. 174) by holding the rubber band at the same tension and moving the movable pencil.

After completing the table, have students write a general rule to describe their observations in terms of distance (length) and pitch.

Table 9.5. Data collection chart for Explore activity for Lesson PS4.A

Distance Between End and Movable Pencil	Description of Sound Heard (Pitch)
25 cm	
20 cm	
15 cm	
20 cm	
25 cm	

Ask the students the following questions:

- What is the difference between the pencil placement for high and low sounds?
- Which will produce the highest sound: three inches, five inches, or seven inches?
- Which will produce the lowest sound: three inches, five inches, or seven inches? How do you know?

Review with students the following questions:

- What is causing the sounds you are hearing? *[Answer: vibrations (sound waves)]*
- What is vibrating? *[Answer: the rubber band, the air particles, and parts of your ear.]*
- Are the sounds alike when you change the placement of the pencil on the rubber band? *[Answer: No.]* How are the sounds different? *[Answer: the pitch]*
- What do you think causes the differences in sound? *[Answer: The shorter the distance between the ends of the rubber band (measured from the movable pencil to one of the secure ends), the higher the pitch. The length is short, so it vibrates many times quickly. The faster it vibrates, the higher the pitch. As the length of the vibrating part gets longer, the vibrating slows down; thus, the pitch gets lower. The longer the vibrating part, the slower the up-and-down movement and the lower the sound produced.]*

Have students respond to the key question in their science notebooks: How does the movement (vibration) of an object affect the sound (pitch)?

ELABORATE

- Describe how students will develop a more sophisticated understanding of the concept.

- What vocabulary will be introduced, and how will it connect to students' observations?

- How is this knowledge applied in our daily lives?

Demonstrate how sounds are produced and controlled in various musical instruments (xylophone, guitar, recorder, piano, etc.). Have students identify the source of the vibrations when the instruments are played.

Ask the following question: What causes the change in pitch on the instruments? Listen to a piece of music for the variance in pitch. Use a "thumbs up" for pitch that goes up and "thumbs down" for pitch that goes down.

Ask the following question: Why would there be a change in pitch throughout a song? [*Answers will vary: adds beauty, pleasure, a sense of movement, variety, etc.*]

Revisit the Engage part of the lab. Identify which waves would be considered high pitch and which waves would be considered low pitch. Have students explain how they identified high-pitch waves and low-pitch waves.

StEMT*ify*

Science: What real-world issues and problems exist in the area of sound energy?

Engineering: Is the STEM lesson guided by the engineering design process?

- Students should be immersed in hands-on inquiry and open-ended exploration.

- Students should be involved in productive teamwork.

Mathematics: Does the STEM lesson apply rigorous mathematics and science content that the students are learning?

Technology: Does the STEM lesson allow for multiple right answers through prototype development, testing, and improvement of design? Encourage students to monitor and evaluate their progress, change course if necessary, and ask, "Does this make sense?"

StEMT*ified* Question: How can a knowledge of wavelength and pitch be used to design musical instruments capable of producing a specific pitch?

Step 1: Ask—*Practice of asking questions (science) and defining problems (engineering)*

What is the problem to solve? What needs to be designed, and who is it for? What are the project requirements and limitations? What is the goal?

Step 2: Research and design—*Practice of planning and carrying out investigations*

Students collaborate to brainstorm ideas and develop as many solutions as possible.

Provide students with an assortment of materials: string or chord, dowel rods, rubber bands, shoe boxes, straws (show students how to cut a reed on the end of the straw), bottles, and water.

Step 3: Plan—*Practice of constructing explanations and designing solutions*

Students will compare the best ideas, select one solution, and make a plan to investigate it.

Students will be given a specific pitch they must match by building a musical instrument out of the provided materials. The original pitch can either be generated by a computer or by using a small inexpensive keyboard or tuning forks.

Step 4: Create—*Practice of developing and using models*

Students will build a prototype.

Step 5: Test and improve—*Practice of obtaining, evaluating, and communicating information*

Does the prototype work, and does it solve the need? Students communicate the results and get feedback, and analyze and talk about what works, what doesn't, and what could be improved.

Look-Fors

Students should engage in the following practices:

- *Practice of planning and carrying out investigations.* Students will be actively engaged and work cooperatively in small groups to complete investigations, test solutions to problems, and draw conclusions. Use rational and logical thought processes, and effective communication skills (writing, speaking, and listening; SEP 7, SEP 8; MP 3).

Figure 9.3. Sample student models

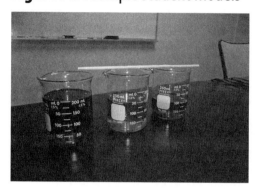 

- *Practice of developing and using models.* Students will obtain, evaluate, and communicate information by constructing explanations and designing solutions (SEP 8; MP 3). Students will develop and use models (musical instrument) to describe the cause-and-effect relationship of pitch and sound (SEP 2; MP 4). (See Figure 9.3 for examples of models that students could make.)

- *Practice of obtaining, evaluating, and communicating information.* Students will analyze and interpret data to draw conclusions and apply understandings to new situations (SEP 4; MP 5). Apply scientific vocabulary after exploring a scientific concept (SEP 6; MP 7).

EVALUATE

- How will students demonstrate that they have achieved the lesson objective?

- Evaluation should be embedded throughout the lesson as well as at the end of the lesson.

Summative Assessment: Write about it!

Students will write a description of their instrument and describe why the instrument is capable of producing the pitch as it relates to wavelength.

StEMT/EDC

Students will develop a model to demonstrate that pitch is a result of the length of a sound wave. Students will use content-specific vocabulary to identify the relationship between pitch and sound using such tools as diagrams, graphs, and the design prototype. Students reflect on whether the results make sense, possibly improving the model if it has not served its purpose (the goal of the desired pitch).

Mirror, Mirror, on the Wall

Physical Sciences: Waves and Their Applications in Technologies for Information Transfer—Electromagnetic Radiation (PS4.B)

Teacher Brief

This lesson is based on the classic misconception about reflections in mirrors with regard to distance. We use this same lesson to demonstrate STEM to our new teachers because it is fun and very effective at demonstrating that students have misconceptions that are difficult to break. It also demonstrates how easy it is to incorporate an engineering design challenge into a lesson. The question "How big does a mirror have to be to see yourself," becomes the logical next question after the conclusion of the initial lesson.

The lesson we have included is a very simple version with many possible variations and additions that could be included depending on grade level. Other options could be to have students line up facing the wall and have each student roll a ball toward a mark on the floor, making predictions of where the ball will end up. The action and reaction of the ball can be recorded in science notebooks. We often use this method, especially in lower grades, to demonstrate how light (using a tennis ball as a proxy) travels in a straight line until it encounters an object. For higher grades, this concept can easily be converted into the law of reflection.

The engineering design challenge uses the question "How big does a mirror have to be for you to see the full height of your body?" Using a long string, students mark the path of a beam of light. Although groups can typically answer the questions through trial and error by separating the two mirrors at different distances, the concept of *how* light travels is often misunderstood if the string or yarn is not used. Make sure students understand the source of light and the direction of travel.

Prior Knowledge

By the end of grade 5. An object can be seen when light reflected from its surface enters the eyes; the color people see depends on the color of the available light sources as well as the properties of the surface. This phenomenon is observed only. The stress is on understanding that light traveling from the object to the eye determines what is seen. Because lenses bend light beams, they can be used, individually or in combination, to provide magnified images of objects too small or too far away to be seen with the naked eye (*Framework;* NRC 2012, p. 135).

Framework Questions: What is light? How can one explain the varied effects that involve light? What other forms of electromagnetic radiation are there?

StEMT*ified* Question: How can we use our knowledge of the reflection of light to design an inexpensive mirror (i.e., how long vertically does a mirror have to be for you to see your entire body)?

Dimension 1: Practices (What should students be doing?)

- Developing and using models
- Constructing explanations and designing solutions
- Engaging in argument from evidence
- Obtaining, evaluating, and communicating information

Dimension 2: Crosscutting Concepts

- *Patterns.* Observed patterns of forms and events guide organization and classification, and they prompt questions about relationships and the factors that influence them.

- *Cause and effect: Mechanism and explanation.* Events have causes—sometimes simple, sometimes multifaceted. A major activity of science is investigating and explaining causal relationships and the mechanisms by which they are mediated. Such mechanisms can then be tested across given contexts and used to predict and explain events in new contexts.

Disciplinary Core Idea

Electromagnetic radiation (e.g., radio, microwaves, light) can be modeled as a wave pattern of changing electric and magnetic fields or, alternatively, as particles. Electromagnetic waves can be detected over a wide range of frequencies, of which the visible spectrum of colors detectable by human eyes is just a small part. Many modern technologies are based on the manipulation of electromagnetic waves (NRC 2012, pp.133–134).

Instructional Duration: Four to five 45-minute class periods

Materials per Class of 30

Two flat mirrors, one spoon, one convex mirror, one concave mirror, yarn (approximately 20 m), simple calculator, safety glasses or goggles

Safety Notes

1. Personal protective equipment (safety glasses or goggles) should be worn during the setup, hands-on, and takedown segments of the activity and demonstration.

2. Use caution when handling sharps (mirror, lenses), as the edges can cut skin.

3. Wash hands with soap and water upon completing this activity.

For a basic overview of physical science, teachers may use the OER CK–12 Standards-Aligned FlexBook textbook, *Physical Science (www.ck12.org/teacher)*.

Guiding Question: What observations help us know how light travels?

Misconceptions

- Students think the object is the source of the light. Correct Understanding: It is the reflection of the light off the object that enables us to see the object.

- Students think that they can see in the complete absence of light. Correct Understanding: Even dark rooms contain some ambient light from the outside (from cars, street lamps, etc.). If there was complete absence of light, you would not be able to see anything. Students think reflection and refraction are the same. Correct Understanding: Reflection is the bouncing of light off a surface; refraction is the bending of light at the contact zone between two mediums, such as air and water.

- Students think refraction occurs continuously through a medium. Correct Understanding: Light only changes direction at the interface between two mediums.

ENGAGE

- How will you capture students' interest and reveal misconceptions?
- What kinds of questions should the students ask themselves after the engagement?

Ask your students the following question: As you back up from a flat mirror, do you see more, less, or the same amount of yourself? Give students 10 seconds of think time to choose an option based on their prior knowledge.

Four Corners Activity (but only three will be used): Place the following signs in three different corners of the room: more, less, and same. Ask students to move to the corner of the answer they have selected and, once in the corner, to choose a partner. Partners discuss their choice and the reason why they have chosen. Teacher walks around and checks for misconceptions and prior knowledge. Do not correct misconceptions.

EXPLORE

- Describe the hands-on activities that will provide experience of the phenomenon.
- List "big idea" conceptual questions you will use to encourage and focus students' exploration and to allow the students to test their ideas.

Students return to their original groups. Distribute team baskets with mirrors. Students use the mirror to answer the engagement question (allow 10–15 minutes). Students write down observations in their log books. Ask probing questions (facilitate; do not direct).

Procedure: One student holds a small flat mirror against the wall so that another student facing the mirror can center his or her face inside the mirror. The student facing the mirror should stand approximately one meter in front of the mirror. The student will then walk slowly backward gathering information with each step. After 10–15 steps, the student will answer the following questions: 1) As you walked away from the mirror, did you ever see your chest, waist, knees, or shoes? 2) As you walked away from the mirror, did you ever see only your eyes or nose? If the answer is no to both questions, and it should be, then in a flat mirror at any distance from the mirror will you always see the same amount of yourself?

EXPLAIN

- Student explanations should precede your explanations or introduction of terms. What questions or techniques will you use to help students connect their exploration to the concept under examination?

- List higher-order thinking questions that you will use to solicit *student* explanations and to help students construct and justify their explanations.

Whip Around: Randomly choose one student from each group to stand. Ask the engaging question. Go around the class, allowing many students to share their answers. Check for understanding and ask clarifying questions.

ELABORATE

- Describe how students will develop a more sophisticated understanding of the concept.

- What vocabulary will be introduced, and how will it connect to students' observations?

- How is this knowledge applied in our daily lives?

Ask students the following question: How would you see yourself in a curved mirror?

Procedure: Distribute a spoon to each student. Students Explore using the spoons (10–15 min) and comparing and contrasting the sides of the spoon as mirrors. Pass out a concave and convex mirror to each group to make sure they have the images correct. Have students record their findings (in terms of the size and orientation of the image) using Table 9.6. *Teacher Note:* You will need to prompt students to place the spoons very close to their eyes so they can see themselves right-side up. This is when the optional convex and concave mirrors could be passed out.

Table 9.6 Data collection chart for Lesson PS4.B

Distance	Concave	Convex
Far away	*Upside down/smaller*	*Upright/larger*
Close up	*Upright/smaller*	*Upright/even smaller yet*

Ask students the following questions:

- Where have you seen these types of mirrors before? *[Answer: Take all answers.]*

- Why is the convex mirror used in hallways near corners? *[Answer: It enables the viewer to have a wide field of view to see someone coming from another hallway.]*

- Where might the concave mirror be useful? *[Answer: The concave mirror magnifies objects if they are close to the mirror (e.g. makeup mirrors).]*

- What type of curved mirror is on the right side of a car? Hint: Think about the words written on the mirror. *[Answer: Images in a convex mirror are always upright. They are smaller, but the advantage comes from the elimination of blind spots with the increased field of view.]*

StEMT*ify*

Science: What real-world issues and problems exist in the area of optics?

Engineering: Is the STEM lesson guided by the engineering design process?

- Students should be immersed in hands-on inquiry and open-ended exploration.

- Students should be involved in productive teamwork.

Mathematics: Does the STEM lesson apply rigorous mathematics and science content that the students are learning?

Technology: Does the STEM lesson allow for multiple right answers through prototype development, testing, and improvement of design? Encourage students to monitor and evaluate their progress, change course if necessary, and ask, "Does this make sense?"

StEMT*ified* Question: How can we use our knowledge about the reflection of light to design an inexpensive mirror? How long vertically does a mirror have to be for you to see your entire body?)

Step 1: Ask—Practice of asking questions (science) and defining problems (engineering)

What is the problem to solve? What needs to be designed, and who is it for? What are the project requirements and limitations? What is the goal?

Parameters of the mirror: The mirror students will construct, working with only two small mirrors, will be as small in vertical length as possible for a person to see their entire body. Assume that the mirror will be 30 centimeters wide. If glass for the mirror costs five cents per square centimeter, what is the cost of the mirror?

- Provide student groups with two flat mirrors (Mirror 1 and Mirror 2) and a length of yarn about 10 meters long.

- Students will use the yarn to illustrate the ray of light that travels between what we see to the mirror and then to our eyes.

Step 2: Research and Design—Practice of planning and carrying out investigations

Students collaborate to brainstorm ideas and develop as many solutions as possible.

Step 3: Plan—Practice of constructing explanations and designing solutions

Students will compare the best ideas, select one solution, and make a plan to investigate it.

Students will place the mirrors against the wall, one above the other. They will stretch the yarn between their eyes, Mirror 1 and what they see. Students will repeat this process with Mirror 2 placed below Mirror 1. With time, students will develop a process to find the minimum size of a mirror as well as its placement on the wall to see their entire body.

Step 4: Create—Practice of developing and using models

Students will design a prototype.

Did every group have the same distance between the mirrors that would correspond to the size of the mirror? Why or why not?

What do you notice about the proposed length of the mirror (should be half the length of the student's height starting at his or her head)? Why doesn't the mirror need to extend all the way to the floor? How does this help to explain what we see?

Teacher Note: The mirror will be half the length of the student's body. The groups' mirror design will vary depending on the height of the person they chose to measure during the investigation. A great thing to point out is that you must maintain the variable used for measuring height (i.e., the same student).

Step 5: Test and improve —Practice of obtaining, evaluating, and communicating information

Does the prototype work and does it solve the need? Students communicate the results and get feedback, and analyze and talk about what works, what doesn't, and what could be improved.

Look-Fors: Students should engage in the following practices and behaviors:

- *Practice of planning and carrying out investigations.* Students will be actively engaged and work cooperatively in small groups to complete investigations, test solutions to problems, and draw conclusions. Use rational and logical thought processes, and effective communication skills (writing, speaking and listening; SEP 7, SEP 8; MP 3).

- *Practice of obtaining, evaluating, and communicating information.* Students will analyze and interpret data to draw conclusions and apply understandings to new situations (SEP 4; MP 5). Apply scientific vocabulary after exploring a scientific concept (SEP 6; MP 7).

EVALUATE

- How will students demonstrate that they have achieved the lesson objective?

- Evaluation should be embedded throughout the lesson as well as at the end of the lesson.

Summative Assessment: Write about it!

For the summative assessment, students will complete a R.A.F.T. (Role, Audience, Format, Task) exercise in which they will pretend to be Nemo, who finds kitchen utensils floating down from a passing cruise ship. He finds the spatula (flat mirror) and the spoons very interesting. He spends many hours investigating his appearance in each. After his new discoveries, he can't wait to tell his friends about his newfound knowledge. He writes a letter to his friends, the turtles, to explain mirrors. Students should use specific examples or evidence from their investigations.

Criteria: Conceptual understanding of how light travels in the context of mirrors.

2 points: Student explanation of how light travels is clearly stated using the context of mirrors.

1 point: Student explanation of how light travels is not related to context of mirrors.

0 points: No understanding of how light travels and the relationship between light and mirrors.

Reference

National Research Council (NRC). 2012. *A framework for K–12 science education: Practices, crosscutting concepts, and core ideas.* Washington, DC: National Academies Press.

CONCLUSION

Schools integrating a STEM program into their curriculum must keep in mind the needs of industry and higher education and focus on application and higher-level thinking. Teaching students to use the engineering design process helps remove boundaries between core subjects, allowing for transfer of knowledge and application of academic skills in science and mathematics.

STEM is not just the integration of technology. It is this difference between science and its application that we feel is the crossroads of STEM. For STEM education to grow and prosper, it needs to become more than integration. We make the case to not abandon everything we have learned about how children learn. If STEM is necessary (what is more typically meant is engineering), then where should engineering be placed within a lesson so it can support a concept of science? It seems to us that an engineering design challenge often fits within the Elaborate portion of the 5E Instructional Model. This might also be the location to place other extension activities, such as reading complex text and writing. It could be the place where a situational science component such as a socioscientific issue might be used to make a lesson more relevant to students. We do not believe that every lesson must include engineering, but we do emphasize the importance of this component to deepen student understanding of a concept.

Many STEM lessons found in the public domain do not exhibit the major characteristics that will ensure student engagement and conceptual learning, including real-world issues or problems, inquiry, rigorous content, and open-ended questions with opportunities for students to reframe failure as a necessary part of learning using the engineering design process. The reason most of the lessons found in the public domain miss the mark of inquiry and conceptual learning is they do not incorporate the driving question that matches the *how* to the *why* of exploring natural phenomena.

Foremost in the case for using the StEMT process for lesson construction, technology is not about what is used within the lesson, but more about a solution to a problem, which is why it is at the end of the acronym *StEMT*. That is not to say that technology should not be embedded throughout every aspect of a STEM lesson, as it should. STEM using the StEMT process is more than integration; it is about using the constructivist learning tools developed throughout decades of research to support the learning of STEM or STEM instruction. The heart of STEM is not engineering but rather problem solving.

Doing STEM does not need to be difficult for teachers. By helping teachers understand the method for StEMT*ifying* their existing lessons and turning them into STEM lessons, engineering becomes another tool in their arsenal that is helpful and beneficial for students. Our goal throughout this process has always been to make STEM more understandable for teachers who can then design effective and meaningful STEM lessons on their own to promote the continued increase in students' scientific literacy that we all desire as outcomes for anyone entering the 21st century.

CONCLUSION

Science (infused with instructional technology tools), engineering, mathematics, and technology (StEMT) is a *process,* and StEMT*ifying* the lesson within the Elaborate section of a 5E lesson, where engineering design and problem solving are used to make relevant meaning of the science concept, is also a process. The implementation of science through inquiry extends to the use of engineering projects within any STEM program, which confronts misconceptions about science concepts. Students who use inquiry to learn science engage in many of the same activities and thinking processes as scientists, and these processes, or practices, are identified in the *Common Core State Standards* and the *Framework for K–12 Science Education.* The practices, whether engineering, science, or mathematics, are the process standards of problem solving, reasoning and evidence, communication, modeling, and applications. Opportunities for students to immerse themselves in these practices and to explore why they are central to mathematics, science, and engineering are critical to the StEMT process.

We hope this book helps school district's curriculum staff to promote STEM as a process for learning, with the purpose being that our students become scientifically literate.

Go **S^T(EM)=>T***ify* those lessons!

APPENDIXES

Appendix A

Claims, Evidence, Reasoning Rubric

Component	Level		
	0	**1**	**2**
Claim—A statement or conclusion that answers the original question or problem.	Does not make a claim or makes an inaccurate claim.	Makes an accurate but incomplete claim.	Makes an accurate and complete claim.
Evidence—Scientific data that support the claim. The data needs to be appropriate and sufficient to support the claim, and observations and measurements are about the natural world.	Does not provide evidence or only provides inappropriate evidence (evidence that does not support the claim).	Provides appropriate but insufficient evidence to support the claim. May include some inappropriate evidence.	Provides appropriate and sufficient evidence to support the claim.
Reasoning—A justification that links the claim to the evidence. It shows why the data count as evidence by using appropriate and sufficient scientific principles (core ideas).	Does not provide reasoning or only provides reasoning that does not link evidence to the claim	Provides reasoning that links the claim and evidence. Repeats the evidence or includes some—but not sufficient—scientific principles.	Provides reasoning that links evidence to the claim. Includes appropriate and sufficient scientific principles.
Rebuttal—Alternative explanations and counterevidence and reasoning for why the alternative explanation is not an appropriate explanation for the question or problem.			

Source: Adapted from K. McNeill and Krajcik, J. (2012). *Supporting Grade 5–8 Students in Constructing Explanations in Science.* New York: Pearson.

Appendix B
Constellations

Orion (Winter Constellation)

Scorpio (Summer Constellation)

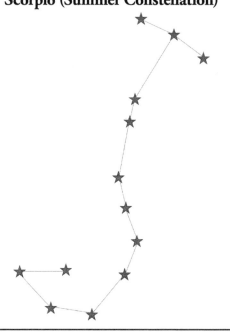

Leo (Spring Constellation)

Pegasus (Fall Constellation)

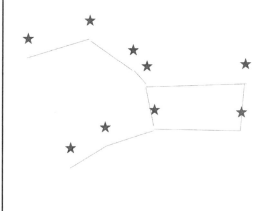

National Science Teachers Association

Appendix C

Moon Phases Model Sheet

Name: _____ Date: _____ Period: _____

Instructions: First, place the ball in the center of the diagram as if it were Earth. Shine the flashlight to represent the Sun's rays. Notice which side of the sphere is lit. Shade in the side that is in shadow. Next, place the sphere on the other blank circles. Again, shine the flashlight on the sphere and notice which side is lit and which side is in shadow. Shade in the appropriate side of the circle that is shaded. Repeat this process until all the blank circles have been shaded.

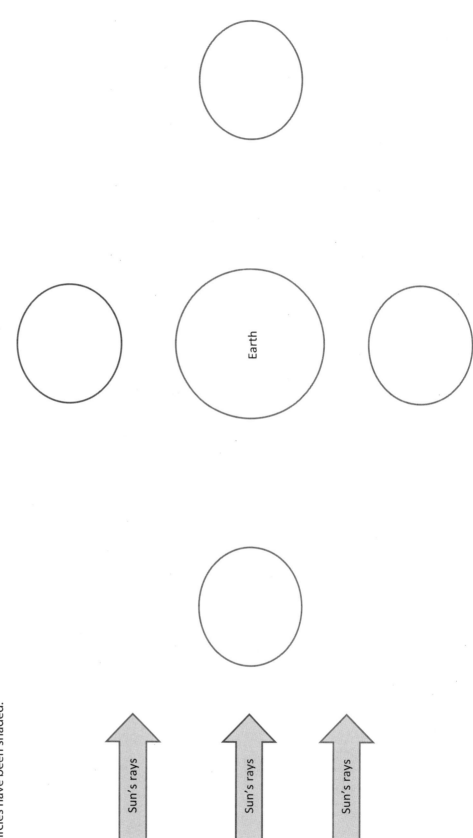

Index

Page numbers printed in **boldface** type refer to figures and tables

INDEX

INDEX

INDEX